DATE			

HEAT AND
MASS TRANSFER
DATA BOOK

HEAT AND MASS TRANSFER DATA BOOK

Third Edition

C. P. KOTHANDARAMAN

S. SUBRAMANYAN

Department of Mechanical Engineering
PSG College of Technology, Coimbatore

A HALSTED PRESS BOOK

JOHN WILEY & SONS
New York London Sydney Toronto

Copyright © 1977, Wiley Eastern Limited

Published in the U.S.A., Canada, Latin America
and the Middle East by Halsted Press,
a Division of John Wiley & Sons, Inc., New York

Library of Congress Cataloging in Publication Data

Kothandaraman, C P
   Heat and mass transfer data book.

   "A Halsted Press book."
   Bibliography : p.
   1. Heat—Transmission—Handbooks, manuals, etc.
2. Mass transfer—Handbooks, manuals, etc. I. Subramanyan, S., joint author. II. Title.
QC 320.4. K67   1977   536'.2   77-7346
ISBN 0-470-99078-3

Printed in India
at Rajbandhu Industrial Company, New Delhi 110 064

FOREWORD

Engineering education will be purposeful only if engineering and basic sciences can be applied to practical situations without the drudgery of memorizing the formulae or data. It is to this end that data books are useful in the teaching-learning process.

The study of heat and mass transfer has been recognized as important and is included as a core subject for students of mechanical engineering in many of our universities.

This compilation of formulae and data completely in metric units in the subject of heat and mass transfer, I hope, will be quite useful for the engineering teachers and students.

It will be very desirable that our technical institutions change over to the use of SI units as rapidly as possible for its elegance and simplicity. MKS and SI units have been used in this book by the compilers with this end in view.

G.R. Damodaran

PSG College of Technology
Coimbatore

PREFACE

The aim of this book is to present to the students, teachers and practising engineers, a comprehensive collection of various material property data and formulae in the field of heat and mass transfer. The material is organized in such a way that a reader who has gone through the engineering curriculum could easily use the formulae and data presented in heat transfer calculations. Hence this compilation is primarily intended as an adjunct to a standard text.

The data book devotes considerable space to the property values of materials—solids, liquids and gases—that are commonly used in heat transfer situations. Property values for various materials at different temperatures are given. This may be found to be of good use to the designers. Also the students can be made to appreciate the influence of variation of property values on the heat transfer process.

The formulae for each chapter are arranged in an easily usable tabular form with symbols and units explained alongside. The limitations and restrictions in the use of empirical relationships are also mentioned alongside. The empirical formulae and charts have been selected with a view to presenting as much information as possible which are simple and straightforward for use without much of mathematical manipulation. Several favourable comments that we received on this point made us to maintain this mode of presentation.

Several suggestions received since the appearance of the second edition have been incorporated, as far as possible, in the new edition. Data on heavy water and some nuclear materials have been added. A number of charts dealing with transient conduction and radiation have been redrawn to improve clarity.

The material presented was collected and used during the past fifteen years while teaching the subject to the undergraduate students at our college. It has served us in training the students with problems of purposeful and practical design situations. The data book has been favourably received by several Indian universities and many of them have permitted its use in the university examinations for reference.

The increasing use of SI Units in several parts of the world prompted us to present the data in the SI Units alongside MKS. This we believe will increase the usefulness of the book.

We acknowledge gratefully the encouragement given by Professor G.R. Damodaran in our effort. Our thanks are due to our principal, Dr. R. Subbayyan for the facilities afforded at the college for the compilation of this handbook. We acknowledge the generous help given by Professor K. Venkataraman, who read the entire manuscript and made many helpful suggestions for improvement.

Coimbatore
January, 1977

C.P. Kothandaraman
S. Subramanyan

CONTENTS

PROPERTY VALUES OF METALS AT 20° C

Metal	Density ρ kg/m³	Thermal Diffusivity α m²/hr	Specific Heat C_p kcal/kg°C	Thermal Conductivity k kcal/m-hr°C	Specific Heat C_p kJ/kg K	Thermal Conductivity k W/mK
Aluminium, pure	2707	0.340	0.214	175.6	0.896	204.2
Al-Cu Duralumin 94-96% Al, 3-5% Cu, trace Mg	2787	0.239	0.211	141.4	0.883	164.5
Al-Si (Silumin Cu Bearing) 86.5% Al, 12.5% Si, 1% Cu	2659	0.215	0.207	117.6	0.867	136.8
Lead	11393	0.086	0.031	29.8	0.130	34.7
Iron, pure	7897	0.073	0.108	62.5	0.452	72.7
Wrought iron < 0.5% C	7849	0.059	0.110	50.6	0.461	58.9
Steel (carbon steel):						
Approx. 0.5 C	7833	0.053	0.111	46.1	0.465	53.6
" 1.0 C	7801	0.042	0.113	37.2	0.473	43.3
" 1.5 C	7753	0.035	0.116	31.2	0.486	36.3
Nickel Steel:						
0% Ni	7897	0.073	0.108	62.5	0.452	72.7
20% Ni	7983	0.019	0.110	16.4	0.461	19.1
40% Ni	8196	0.010	0.110	8.9	0.461	10.4
80% Ni	8618	0.032	0.110	29.8	0.461	34.7
Invar 36% Ni	8137	0.010	0.110	9.2	0.461	10.7
Chrome Steel:						
0% Cr	7897	0.073	0.108	62.5	0.452	72.7
1% Cr	7865	0.060	0.110	52.1	0.461	60.6
5% Cr	7833	0.040	0.110	34.2	0.461	39.8
20% Cr	7689	0.024	0.110	19.3	0.461	22.5
Chrome Nickel:						
15% Cr, 10% Ni	7865	0.019	0.110	16.4	0.461	19.1
18% Cr, 8% Ni (V₂A)	7817	0.016	0.110	14.0	0.461	16.3
25% Cr, 20% Ni	7865	0.013	0.110	11.0	0.461	12.8

1

PROPERTY VALUES OF METALS AT 20° C (Cont.)

Metal	Density ρ kg/m³	Thermal Diffusivity α m²/hr	Specific Heat C_p kcal/kg°C	Thermal Conductivity k kcal/m-hr°C	Specific Heat C_p kJ/kg K	Thermal Conductivity k W/mK
German Silver 62% Cu, 15% Ni, 22% Zn	8618	0.027	0.094	21.4	0.394	24.9
Constantan 60% Cu, 40% Ni	8922	0.022	0.098	19.5	0.410	22.7
Magnesium, pure	1746	0.349	0.242	147.3	1.013	171.3
Mg-Al (Electrolytic) 6-8% Al, 1-2% Zn	1810	0.129	0.240	56.5	1.005	65.7
Molybdenum	10220	0.192	0.060	105.6	0.251	122.8
Nickel, pure (99.9%)	8906	0.082	0.106	77.4	0.444	90.0
Nickel Chrome 90% Ni, 10% Cr	8666	0.016	0.106	14.8	0.444	17.2
Tungsten Steel:						
0% W	7897	0.073	0.108	62.5	0.452	72.7
1% W	7913	0.067	0.107	56.5	0.448	65.7
5% W	8073	0.055	0.104	46.1	0.435	53.6
10% W	8313	0.049	0.100	41.7	0.419	48.5
Copper, pure	8954	0.404	0.091	331.9	0.381	386.0
Aluminium Bronze 95% Cu, 5% Al	8666	0.084	0.098	71.4	0.410	83.0
Bronze 75% Cu, 25% Sn	8666	0.031	0.082	22.3	0.343	25.9
Red Brass 85% Cu, 9% Sn, 6% Zn	8714	0.065	0.092	52.1	0.385	60.6
Brass 70% Cu, 30% Zn	8522	0.122	0.092	95.2	0.385	110.7
Silver, pure	10524	0.596	0.056	349.8	0.235	406.8
Tungsten	19350	0.226	0.032	139.9	0.134	162.7
Zinc, pure	7144	0.148	0.092	96.4	0.385	112.1
Tin, pure	7304	0.139	0.054	55.1	0.226	64.1

PROPERTIES OF ELEMENTS AT 0°C

Element	Density ρ kg/m³	Thermal Diffusivity α m²/hr	Specific Heat C_p kcal/kg°C	Thermal Conductivity k kcal/m-hr°C	Specific Heat C_p kJ/kg K	Thermal Conductivity k W/mK
Beryllium	1840	0.1835	0.4	135	1.675	157
Cadmium	8660	0.168	0.0549	80	0.2299	93
Carbon (graphite)	1700 to 2300	0.440	0.16	100 to 150	0.67	116.3 to 174.5
Chromium	7150	0.0785	0.107	60	0.448	69.8
Cobalt	8800	0.0635	0.107	60	0.448	69.8
Lithium	5340	0.145	0.79	59	3.308	68.6
Molybdenum	10200	0.196	0.0603	121	0.2525	140.7
Platinum	21460	0.0885	0.0316	60	0.1323	69.8
Potassium	870	0.560	0.176	86	0.7369	100.0
Sodium	975	0.340	0.286	94	1.1974	109.3
Uranium	19100	0.0309	0.028	16.5	0.1172	19.9
Vanadium	5900	0.043	0.1185	30	0.4961	34.9
Titanium	4540	0.0224	0.127	13	0.5317	15.12

VARIATION OF THERMAL CONDUCTIVITY OF METALS WITH TEMPERATURE (kcal/m-hr°C)

Metal \ Temperature °C	-100	0	100	200	300	400	600	800	1000	1200
Aluminium, pure	184.5	174.1	177.1	184.5	196.5	214.3	-	-	-	-
Al-Cu Duralumin 94-96% Al, 3-5% Cu, Trace Mg	108.6	136.9	156.3	166.7	-	-	-	-	-	-
Al-Si (Silumin Cu Bearing) 86.5% Al, 12.5% Si, 1% Cu	102.6	117.6	123.5	130.9	138.4	-	-	-	-	-
Lead	31.7	30.2	28.7	27.0	25.6	-	-	-	-	-
Iron, pure	74.4	62.5	58.0	53.6	47.6	41.7	34.2	31.2	29.8	31.2
Wrought Iron < 0.5% C	-	50.6	49.1	44.6	41.7	38.7	31.2	28.3	28.3	28.3
Steel (Carbon Steel):										
C ≈ 0.5%	-	47.6	44.6	41.7	38.7	35.7	29.8	26.8	25.3	26.8
1.0%	-	37.2	37.2	35.7	34.2	31.2	28.3	25.3	23.8	25.3
1.5%	-	31.2	31.2	31.2	29.8	28.3	26.8	23.8	23.8	25.3
Chrome Steel:										
0% Cr	74.4	62.5	58.0	53.6	47.6	41.7	34.2	31.2	29.8	31.2
1% Cr	-	53.6	47.6	44.6	40.2	35.7	31.2	28.3	28.3	-
5% Cr	-	34.2	32.7	31.2	31.2	28.3	25.3	25.3	25.3	25.3
20% Cr	-	19.3	19.3	19.3	19.3	20.8	20.8	22.3	25.3	-
Chrome Nickel : 18% Cr, 8% Ni, (V2A)	-	14.0	14.9	14.9	16.4	16.4	19.3	22.3	26.8	-
German Silver : 62% Cu, 15% Ni, 22% Zn	16.5	-	26.8	34.2	38.7	41.7	-	-	-	-
Constantan : 60% Cu, 40% Ni	17.9	-	19.0	22.3	-	-	-	-	-	-

4

VARIATION OF THERMAL CONDUCTIVITY OF METALS WITH TEMPERATURE (kcal/m·hr°C) (Cont.)

Metal \ Temperature °C	-100	0	100	200	300	400	600	800	1000	1200
Magnesium, pure	153.3	147.3	144.3	140.0	135.4	-	-	-	-	-
Mg-Al (Electrolytic) 6-8% Al, 1-2% Zn	-	44.7	53.6	64.0	71.4	-	-	-	-	-
Molybdenum	119.0	107.1	101.2	98.3	95.3	93.8	90.8	87.8	84.8	78.9
Nickel, pure (99.9%)	89.3	80.4	71.5	62.5	55.1	50.6	-	-	-	-
Nickel Chrome 90% Ni, 10% Cr	-	14.7	16.2	18.0	19.7	21.2	-	-	-	-
Copper, pure	349.7	331.9	325.9	321.5	317.0	312.6	303.6	-	-	-
Red Brass 85% Cu, 9% Sn, 6% Zn	-	50.6	61.0	-	-	-	-	-	-	-
Brass : 70% Cu, 30% Zn	75.9	-	110.1	123.7	126.5	126.5	-	-	-	-
Silver, pure (99.9%)	360.2	359.0	358.0	321.5	311.5	310.0	-	-	-	-
Tungsten	-	142.9	129.5	122.0	114.7	108.7	96.7	65.5	-	-
Zinc, pure	98.3	96.7	93.8	90.8	86.4	80.4	-	-	-	-
Tin, pure	64.0	56.7	50.6	49.1	-	-	-	-	-	-
Bronze : 57% Cu, 37% Zn, 2.3% Mn	-	60.0	63.0	67.0	71.0	76.0	-	-	-	-
Uranium	-	16.5	17.5	20.0	20.8	-	-	-	-	-
Zirconium	-	-	18.0	17.5	17.1	17.1	18.5	-	-	-
Zirconium 97%, and Tin 3%	-	-	10.3	11.4	12.5	13.4	15.5	-	-	-

VARIATION OF THERMAL CONDUCTIVITY OF METALS WITH TEMPERATURE (W/mK)

Metal \ Temperature °C	-100	0	100	200	300	400	600	800	1000	1200
Aluminium, pure	214.6	202.5	206.0	214.6	228.5	249.2	-	-	-	-
Al-Cu Duralumin 94-96% Al, 3-5% Cu, Trace Mg	126.5	159.2	181.8	193.9	-	-	-	-	-	-
Al-Si (Silumin Cu Bearing) 86.5%Al 12.5%Si 1%Cu	119.3	136.8	143.6	152.2	161.0	-	-	-	-	-
Lead	36.9	35.1	33.4	31.4	29.8	-	-	-	-	-
Iron, pure	86.5	72.7	67.5	62.3	55.4	48.5	39.8	36.3	34.7	36.3
Wrought Iron C < 0.5%	-	58.9	57.1	51.9	48.5	45.0	36.3	32.9	32.9	32.9
Steel (Carbon Steel) approx. C 0.5%	-	55.4	51.9	48.5	45.0	41.5	34.7	31.2	29.4	31.2
" 1.0%	-	43.3	43.3	41.5	39.8	36.3	32.9	29.4	27.7	29.4
" 1.5%	-	36.3	36.3	36.3	34.7	32.9	31.2	27.7	27.7	29.4
Chrome Steel Cr 0%	86.5	72.7	67.5	62.3	55.4	48.5	39.8	36.3	34.7	36.3
1%	-	62.3	55.4	51.9	46.8	41.5	36.3	32.9	32.9	-
5%	-	39.8	38.0	36.3	36.3	32.9	29.4	29.4	29.4	29.4
20%	-	22.4	22.4	22.4	22.4	24.2	24.2	25.9	29.4	-
Chrome Nickel 18% Cr, 8% Ni (V2A)	-	16.3	17.3	17.3	19.1	19.1	22.5	25.9	31.2	-
German Silver 62% Cu, 15% Ni, 22% Zn	19.2	-	31.2	39.8	45.0	48.5	-	-	-	-
Constantan 60% Cu, 40% Ni	20.8	-	22.1	25.9	-	-	-	-	-	-

VARIATION OF THERMAL CONDUCTIVITY OF METALS WITH TEMPERATURE (W/mK) (Cont.)

Metal \ Temperature °C	-100	0	100	200	300	400	600	800	1000	1200
Magnesium, pure	178.3	171.3	167.8	162.8	157.5	-	-	-	-	-
Mg-Al (Electrolytic) 6-8% Al, 1-2% Zn	-	52.0	62.3	74.4	83.2	-	-	-	-	-
Molybdenum	138.4	124.6	117.5	114.3	111.0	109.1	105.6	102.1	98.6	91.8
Nickel, pure (99.9%)	103.9	93.5	83.2	72.7	64.1	58.9	-	-	-	-
Nickel Chrome 90% Ni, 10% Cr	-	17.1	18.8	20.9	22.9	24.7	-	-	-	-
Copper, pure	406.7	386.0	379.0	373.9	368.7	363.6	353.1	-	-	-
Red Brass 85% Cu, 9% Sn, 6% Zn	-	58.9	70.9	-	-	-	-	-	-	-
Brass : 70% Cu, 30% Zn	88.3	-	128.1	143.9	147.1	147.1	-	-	-	-
Silver, pure (99.9%)	418.9	417.5	416.4	373.9	362.3	360.5	-	-	-	-
Tungsten	-	166.2	150.6	141.9	133.4	126.4	112.5	76.2	-	-
Zinc, pure	114.3	112.5	109.1	105.6	100.5	93.5	-	-	-	-
Tin, pure	74.4	65.9	58.9	57.1	-	-	-	-	-	-
Bronze : 57% Cu, 31.1% Zn, 2.3% Mn	69.78	73.62	73.63	77.92	82.57	88.38	-	-	-	-
Uranium	-	19.18	20.35	23.26	24.19	-	-	-	-	-
Zirconium	-	-	20.93	20.35	19.89	19.89	21.52	-	-	-
Zirconium 97%, and Tin 3%	-	-	11.98	13.26	15.54	15.58	18.02	-	-	-

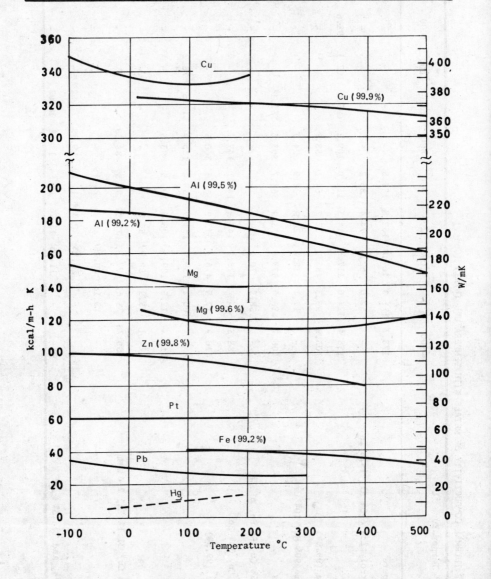

PROPERTY VALUES OF INSULATING, BUILDING AND OTHER MATERIALS

Material	Density ρ kg/m³	Temperature t °C	Thermal Diffusivity α x 10³ m²/hr	Thermal Conductivity k x 10³ kcal/m-hr°C	Specific Heat Cp kcal/kg°C	Thermal Conductivity k x 10³ W/mK	Specific Heat Cp kJ/kg K
Aluminium foil	20	50	-	40	-	46.5	-
Asbestos, fibre	470	50	1.04	95	0.195	110.5	0.816
Asphalt	2110	20	0.57	600	0.50	697.8	2.093
Balsa wood	128	30	-	45	-	52.3	-
Boiler scale	-	65	-	i 130-2 700	-	1 314-3 140	-
Brick, masonry	800-1 500	20	-	200-250	-	233-291	-
Carborundum brick	1 000	-	6.00	9 700	0.162	11 281	0.678
Cardboard, corrugated	-	-	-	55	-	64	-
Cellotex	215	20	-	40	-	46.5	-
Celluloid	1 400	30	-	180	-	209.3	-
Chalk	2 000	50	1.91	800	0.21	930.4	0.879
Clinker	1 400	30	0.41	140	0.34	162.8	1.675
Coal	1 400	20	0.37	160	0.312	186.1	1.306
Coke, powdered	449	100	0.13	164	0.29	190.7	1.214
Concrete	2 300	20	1.77	1 100	0.27	1 279	1.130
Cork, granulated	45	20	-	33	-	38.4	-
Cork, plate	190	30	0.42	36	0.45	41.9	1.884
Earth, dry	1 500	-	-	119	-	138.4	-
Earth, wet	1 700	-	0.69	565	0.48	657.1	2.010
Fibre, plate	240	20	-	42	-	48.9	-
Fibre brick	550	100	-	120	-	139.6	-

PROPERTY VALUES OF INSULATING, BUILDING AND OTHER MATERIALS (Cont.)

Material	Density ρ kg/m³	Tempera-ture t °C	Thermal Diffusivity α x 10³ m²/hr	Thermal Conductivity k x 10³ kcal/m-hr°C	Specific Heat C_p kcal/kg°C	Thermal Conductivity k x 10³ W/mK	Specific Heat C_p kJ/kg K
Glass	2 500	20	1.60	640	0.16	744.3	0.67
Glass wool	200	20	1.00	32	0.16	37.2	0.67
Gravel	1 840	20	-	310	-	360.5	-
Gypsum	1 650	-	-	250	-	290.8	-
Ice	920	0	3.89	1 935	0.54	2 250	2.261
Ice	-	-95	-	3 400	0.28	3 954	1.172
Lamp black	190	40	-	127	-	147.7	-
Leather (sole)	1 000	30	-	137	-	159.3	-
Linoleum	1 180	20	-	160	-	186.1	-
Magnesia, 85% powdered	216	100	-	58	-	67.5	-
Marble	2 700	90	4.15	1 120	1.0	1 303	4.187
Mica	290	-	8.20	500	2.1	581.5	8.792
Mineral wool	200	50	0.91	40	0.22	46.5	0.921
Oak, across grain	800	20	0.53	178	0.42	207.0	1.759
Oak, along grain	800	20	-	312	-	362.9	-
Paraffin	920	20	-	230	-	267.5	-
Peat, plate	220	50	-	55	-	64.0	-
Pine, across grain	448	20	-	92	-	107.0	-
Pine, along grain	448	20	-	220	-	255.9	-
Plaster	1 680	20	-	670	-	779.2	-
Porcelein	2 400	95	1.43	890	0.26	1 035	1.089

PROPERTY VALUES OF INSULATING, BUILDING AND OTHER MATERIALS (Cont.)

Material	Density ρ kg/m³	Temperature t °C	Thermal Diffusivity α x 10³ m²/hr	Thermal Conductivity k x 10³ kcal/m-hr°C	Specific Heat C_p kcal/kg°C	Thermal Conductivity k x 10³ W/mK	Specific Heat C_p kJ/kg K
Porcelein	2 400	1 055	-	1 690	-	1 965	-
Portland cement	1 900	30	0.51	260	0.27	302.4	1.130
Quartz, across grain	2 500-2 800	0	12.00	6 200	0.20	7 211	0.837
Quartz, along grain	2 500-2 800	0	-	11 700	-	13 607	-
Refractory clay	1 845	450	1.86	890	0.26	1 035	1.089
Rubber	1 200	0	0.35	140	0.33	162.8	1.382
Sand, dry	1 500	20	9.85	280	0.19	325.6	0.796
Sand, damp	1 650	20	1.77	970	0.50	1 128	2.093
Saw dust	200	20	-	60	-	69.8	-
Sheet asbestos	770	30	0.71	100	0.195	116.3	0.816
Slag concrete, lumps	2 150	-	1.78	800	0.21	930.4	0.879
Slag wool	250	100	-	60	-	69.8	-
Slate	2 800	100	-	1 280	-	1 489	-
Snow	560	-	1.43	400	0.50	465.2	2.093
Sugar, granulated	1 600	0	1.00	500	0.30	581.5	1.256
Wool felt	330	30	-	45	-	52.3	-

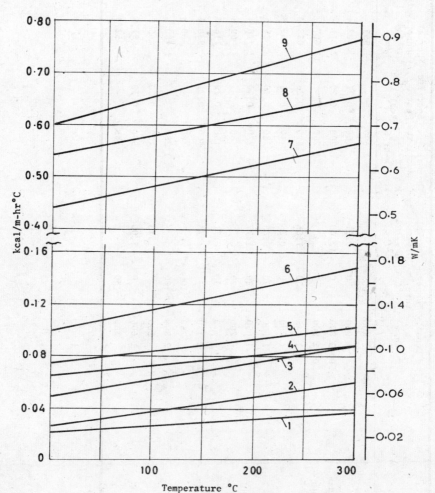

1 Air
2 Mineral wool
3 Slag wool
4 85% Magnesia
5 Sovelite

6 Diatomaceous earth brick
7 Red brick
8 Slag concrete brick
9 Fire clay brick

PROPERTY VALUES OF LIQUIDS IN SATURATED STATE

Temperature °C	Density ρ kg/m³	Kinematic Viscosity ν x 10⁶ m²/s	Thermal Diffusivity α x 10⁶ m²/hr	Prandtl Number Pr	Thermal Conductivity k x 10³ kcal/m-hr°C	Specific Heat C_p kcal/kg°C	Thermal Conductivity k x 10³ W/mK	Specific Heat C_p kJ/kg K
WATER								
0	1002	1.788	471	13.600	475	1.007	552.4	4.216
20	1000	1.006	515	7.020	514	0.998	597.8	4.178
40	995	0.657	544	4.340	540	0.998	628.0	4.178
60	985	0.478	559	3.020	560	0.999	651.3	4.183
80	974	0.364	589	2.220	575	1.002	668.7	4.195
100	961	0.293	605	1.740	585	1.007	680.4	4.216
120	945	0.247	615	1.446	589	1.015	685.0	4.250
140	928	0.213	621	1.241	588	1.023	683.8	4.283
160	909	0.189	622	1.099	585	1.037	680.4	4.342
180	889	0.173	621	1.044	581	1.055	675.7	4.417
200	867	0.160	614	0.937	572	1.076	665.2	4.505
220	842	0.149	605	0.891	561	1.101	652.4	4.610
240	815	0.143	590	0.871	546	1.136	635.0	4.756
260	786	0.137	568	0.874	525	1.182	610.6	4.949
280	752	0.135	533	0.910	499	1.244	580.3	5.208
300	714	0.135	477	1.019	464	1.368	539.6	5.728
AMMONIA								
-50	704	0.435	627	2.600	470	1.066	546.6	4.463
-40	691	0.406	639	2.280	470	1.067	546.6	4.467
-30	679	0.387	648	2.150	472	1.069	548.9	4.476
-20	667	0.381	655	2.090	470	1.077	546.6	4.509
-10	653	0.378	657	2.070	467	1.090	543.1	4.564
0	640	0.373	655	2.050	464	1.107	539.6	4.635
10	626	0.368	648	2.040	457	1.126	531.5	4.714
20	612	0.358	639	2.020	448	1.146	521.0	4.798
30	596	0.350	627	2.010	436	1.168	507.1	4.890
40	581	0.340	612	2.000	424	1.194	493.1	4.999
50	561	0.330	596	1.990	409	1.222	475.7	5.116

PROPERTY VALUES OF LIQUIDS IN SATURATED STATE (Cont.)

Temperature °C	Density ρ kg/m³	Kinematic Viscosity $\nu \times 10^6$ m²/s	Thermal Diffusivity $\alpha \times 10^6$ m²/hr	Prandtl Number Pr	Thermal Conductivity $k \times 10^3$ kcal/m-hr°C	Specific Heat C_p kcal/kg°C	Thermal Conductivity $K \times 10^3$ W/m K	Specific Heat C_p kJ/kg K
CARBON-DIOXIDE								
-50	1156	0.119	145	2.960	73.5	0.440	85.5	1.842
-40	1120	0.118	173	2.460	86.9	0.450	101.1	1.884
-30	1077	0.117	190	2.220	96.0	0.470	111.7	1.968
-20	1032	0.115	196	2.120	98.9	0.490	115.0	2.052
-10	983	0.113	185	2.200	94.5	0.520	109.9	2.177
0	927	0.108	165	2.380	89.9	0.590	104.6	2.470
10	863	0.102	130	2.800	83.5	0.750	97.1	3.140
20	772	0.091	79.9	4.100	75.0	1.200	87.2	5.024
30	598	0.079	10.0	28.700	60.4	8.700	70.3	36.425
GLYCERINE - C₃H₅(OH)₃								
0	1276	8314	354	84700	243	0.540	282.6	2.261
10	1270	3000	347	31000	244	0.554	283.8	2.320
20	1264	1180	341	12500	246	0.570	286.1	2.387
30	1258	501	334	5380	246	0.584	286.1	2.445
40	1252	223	329	2450	246	0.600	286.1	2.512
50	1245	149	321	1630	247	0.617	287.3	2.583
SULPHUR DIOXIDE								
-50	1561	0.484	411	4.240	208	0.325	241.9	1.361
-40	1537	0.423	407	3.740	202	0.325	234.9	1.361
-30	1513	0.370	403	3.310	198	0.325	230.3	1.361
-20	1488	0.324	399	2.930	194	0.325	225.6	1.361
-10	1463	0.288	395	2.620	188	0.325	218.6	1.361
0	1438	0.257	389	2.380	182	0.326	211.7	1.365
10	1412	0.232	384	2.180	176	0.326	204.7	1.365
20	1386	0.209	378	2.000	171	0.326	198.9	1.365
30	1359	0.189	373	1.830	165	0.326	191.9	1.365
40	1329	0.173	367	1.700	159	0.327	184.9	1.369
50	1299	0.162	360	1.610	152	0.327	176.8	1.369

14

PROPERTY VALUES OF LIQUIDS IN SATURATED STATE (Cont.)

Temperature °C	Density ρ kg/m³	Kinematic Viscosity ν x 10⁶ m²/s	Thermal Diffusivity α x 10⁶ m²/hr	Prandtl Number Pr	Thermal Conductivity k x 10³ kcal/m-hr°C	Specific Heat C_p kcal/kg°C	Thermal Conductivity k x 10³ W/m K	Specific Heat C_p kJ/kg K
METHYL CHLORIDE (REFRIGERANT-40) - CH_3Cl								
-50	1052	0.319	500	2.310	185	0.352	215.2	1.474
-40	1033	0.317	492	2.320	180	0.354	209.3	1.482
-30	1016	0.314	481	2.350	174	0.356	202.4	1.491
-20	999	0.309	468	2.380	168	0.359	195.4	1.503
-10	981	0.305	452	2.430	161	0.363	187.2	1.520
0	962	0.302	437	2.490	153	0.367	177.9	1.537
10	942	0.293	420	2.550	147	0.373	171.0	1.562
20	923	0.292	400	2.630	140	0.379	162.8	1.587
30	903	0.288	381	2.720	132	0.386	153.5	1.616
40	883	0.281	359	2.830	124	0.394	144.2	1.650
50	861	0.274	333	2.970	115	0.403	133.8	1.687
DICHLORO DIFLURO METHANE (REFRIGERANT-12) - CCl_2F_2								
-50	1547	0.310	180	6.200	58.0	0.209	67.5	0.875
-40	1518	0.279	185	5.400	59.5	0.211	69.2	0.883
-30	1490	0.253	190	4.800	59.5	0.214	69.2	0.896
-20	1460	0.235	194	4.400	61.0	0.217	70.9	0.909
-10	1429	0.221	198	4.000	62.5	0.219	72.7	0.917
0	1397	0.213	201	3.800	62.5	0.223	72.7	0.934
10	1364	0.203	202	3.600	62.5	0.227	72.7	0.950
20	1330	0.198	202	3.500	62.5	0.230	72.7	0.963
30	1295	0.194	202	3.500	61.0	0.235	70.9	0.984
40	1257	0.191	200	3.500	59.5	0.240	69.2	1.005
50	1216	0.189	196	3.500	58.6	0.244	68.2	1.022
EUTECTIC CALCIUM CHLORIDE SOLUTION (29.9% $CaCl_2$)								
-50	1320	36.352	420	312.000	345	0.623	401.2	2.608
-40	1314	24.971	432	208.000	357	0.629	415.2	2.634
-30	1310	17.177	444	139.000	369	0.636	429.2	2.663

PROPERTY VALUES OF LIQUIDS IN SATURATED STATE (Cont.)

Temperature °C	Density ρ kg/m³	Kinematic Viscosity ν x 10⁶ m²/s	Thermal Diffusivity α x 10⁶ m²/hr	Prandtl Number Pr	Thermal Conductivity k x 10³ kcal/m-hr°C	Specific Heat C_p kcal/kg°C	Thermal Conductivity k x 10³ W/m K	Specific Heat C_p kJ/kg K
-20	1305	11.036	456	87.100	383	0.642	445.4	2.688
-10	1300	6.956	468	53.600	394	0.648	458.2	2.713
0	1296	4.394	479	33.000	406	0.654	472.2	2.738
10	1291	3.353	491	24.600	417	0.660	485.0	2.763
20	1286	2.722	502	19.600	429	0.666	498.9	2.788
30	1282	2.267	512	16.000	439	0.672	510.6	2.813
40	1277	1.923	520	13.300	450	0.678	523.4	2.839
50	1272	1.654	529	11.300	460	0.685	535.0	2.868
ETHYLENE GLYCOL - $C_2H_4(OH)_2$								
0	1131	57.523	336	615	208	0.548	241.9	2.294
20	1116	19.174	338	204	214	0.569	248.9	2.382
40	1101	8.686	338	93	220	0.591	255.9	2.474
60	1087	4.747	335	51	223	0.612	259.4	2.562
80	1077	2.982	332	32.4	225	0.633	261.7	2.650
100	1058	2.025	327	22.4	226	0.655	262.8	2.742
ENGINE OIL (UNUSED)								
0	899	4282	328	47100	127	0.429	147.7	1.796
20	888	901	314	10400	125	0.449	145.4	1.880
40	876	241	300	2870	124	0.469	144.2	1.964
60	864	83	288	1050	121	0.489	140.7	2.047
80	852	37	277	490	119	0.509	138.4	2.131
100	840	20	266	276	118	0.530	137.2	2.219
120	828	12	256	175	116	0.551	134.9	2.307
140	816	7	247	116	115	0.572	133.8	2.395
160	805	5	239	84	113	0.593	131.4	2.483

PROPERTIES OF HEAVY WATER (D_2O):

Molecular weight	: 20.033	Critical Temperature	: 371.5°C
Freezing point	: 3.82°C	Critical Pressure	: 218.6 kgf/cm^2
Boiling point at 1 kgf/cm^2	: 101.43°C		

SPECIFIC HEAT

Temperature °C	Specific heat C_p kcal/kg°C	Specific heat C_p kJ/kg°C
10	1.009	4.226
15	1.007	4.218
20	1.006	4.211
25	1.005	4.207
30	1.004	4.205
35	1.003	4.201
40	1.003	4.201
45	1.003	4.201
50	1.004	4.203

ABSOLUTE VISCOSITY μ

Temperature °C	Viscosity of heavy water (D_2O) / Viscosity of light water (H_2O)
5	1.309
10	1.286
15	1.267
20	1.249
25	1.232
30	1.215
80	1.164
100	1.154
125	1.146

ENTHALPY OF EVAPORATION

Temperature °C	Enthalpy of evaporation kcal/kg	Enthalpy of evaporation kJ/kg	r*D_2O / r H_2O
3.82	554.5	2321.7	1.038
10	549.3	2299.9	1.034
25	541.4	2266.7	1.031
40	531.3	2224.7	1.028
60	519.7	2176.0	1.026
80	507.8	2126.3	1.023
100	495.5	2074.7	1.021
120	482.6	2020.7	1.020
140	465.9	1950.9	1.018
160	454.6	1903.3	1.016
180	439.1	1838.7	1.015
200	422.6	1769.4	1.014
220	404.5	1693.7	1.013

*r - enthalpy of evaporation

COEFFICIENT OF CUBICAL EXPANSION, β

Fluid	Coefficient of cubical expansion at 20°C $\beta \times 10^6$, °K^{-1}
Water	210
Ammonia	2460
Carbon dioxide	14000
Glycerine	504
Sulphur dioxide	1950
Ethylene glycol	648
Engine oil	702
Dichloro difluoro methane (Refrigerant-12) CCl_2F_2	2630

Temperature °C	AIR $\beta \times 10^3$, 1/K	WATER $\beta \times 10^3$, 1/K
5	3.60	0.07
10	3.54	0.11
20	3.42	0.21
30	3.30	0.31
40	3.19	0.41
50	3.10	0.48
60	3.00	0.53
70	2.91	0.58
80	2.83	0.64
90	2.75	0.70
100	2.68	0.76
150	2.36	1.01
200	2.11	1.32
250	1.91	2.92
300	1.75	3.54
500	1.29	-
1000	0.77	-
2000	0.44	-

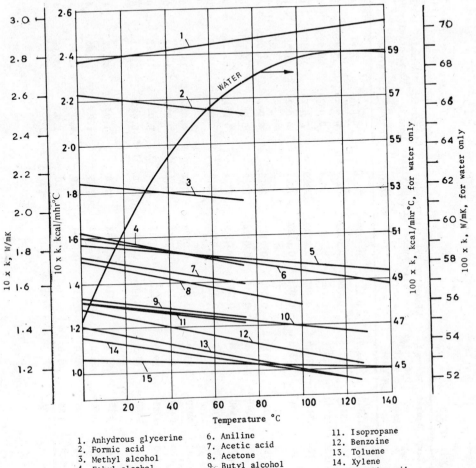

Temperature °C

1. Anhydrous glycerine
2. Formic acid
3. Methyl alcohol
4. Ethyl alcohol
5. Castor oil

6. Aniline
7. Acetic acid
8. Acetone
9. Butyl alcohol
10. Nitrobenzene

11. Isopropane
12. Benzoine
13. Toluene
14. Xylene
15. Vaseline oil

PHYSICAL PROPERTIES OF LIQUID METALS

Name of metal, melting and boiling points	Temp. t °C	Density ρ kg/m³	Thermal Diffusivity a x 10³ m²/hr	Kinematic Viscosity ν x 10⁶ m²/s	Prandtl Number Pr	Thermal Conductivity k kcal/m-hr°C	Specific Heat C_p kcal/kg°C	Thermal Conductivity k W/m K	Specific Heat C_p kJ/kg K
MERCURY, Hg	20	13550	15.7	0.114	0.0272	6.8	0.0332	7.91	0.1390
	100	13350	17.6	0.094	0.0192	7.7	0.0328	8.96	0.1373
t_m = -38.9°C	150	13230	19.1	0.086	0.0162	8.3	0.0328	9.65	0.1373
t_b = 357°C	200	13120	20.6	0.080	0.0140	8.9	0.0328	10.35	0.1373
	300	12880	23.9	0.071	0.0107	10.1	0.0328	11.75	0.1373
TIN, Sn	250	6980	69.0	0.270	0.0141	29.3	0.061	34.08	0.2554
	300	6940	68.5	0.240	0.0126	29.0	0.061	33.73	0.2554
t_m = 231.9°C	400	6865	68.0	0.200	0.0106	28.5	0.061	33.15	0.2554
t_b = 2270°C	500	6790	67.5	0.173	0.0092	28.0	0.061	32.56	0.2554
BISMUTH, Bi	300	10030	31.0	0.171	0.0198	11.2	0.036	13.03	0.1507
	400	9910	35.0	0.142	0.0146	12.4	0.036	14.42	0.1507
t_m = 271°C	500	9785	39.0	0.122	0.0113	13.6	0.036	15.82	0.1507
t_b = 1490°C	600	9660	43.0	0.108	0.0091	14.8	0.036	17.21	0.1507
LITHIUM, Li	200	515	62.0	1.110	0.0643	32.0	1.0	37.22	4.1868
	300	505	66.0	0.927	0.0503	33.5	1.0	38.96	4.1868
t_m = 186°C	400	495	73.0	0.817	0.0404	36.0	1.0	41.87	4.1868
t_b = 1317°C	500	484	81.0	0.734	0.0328	39.0	1.0	45.36	4.1868

PHYSICAL PROPERTIES OF LIQUID METALS (Cont.)

Name of metal, melting and boiling points.	Temp. t °C	Density ρ kg/m³	Thermal Diffusivity $\alpha \times 10^3$ m²/hr	Kinematic Viscosity $\nu \times 10^6$ m²/s	Prandtl Number Pr	Thermal Conductivity k kcal/m-hr°C	Specific Heat C_p kcal/kg°C	Thermal Conductivity k W/m K	Specific Heat C_p kJ/kg K
SODIUM, Na $t_m = 97.3°C$ $t_b = 878°C$	150	916	246	0.594	0.0087	73.0	0.324	84.90	1.3565
	200	903	244	0.506	0.0075	70.0	0.317	81.41	1.3272
	300	878	227	0.394	0.0063	61.0	0.306	70.94	1.2812
	400	854	212	0.330	0.0056	55.0	0.304	63.97	1.2728
	500	829	195	0.289	0.0053	49.0	0.304	56.99	1.2728
ALLOY 56.5% Bi + 43.5% Pb $t_m = 123.5°C$ $t_b = 1600°C$	150	10550	23	0.289	0.0450	8.4	0.035	9.77	0.1465
	200	10490	24	0.243	0.0364	8.9	0.035	10.35	0.1465
	300	10360	27	0.187	0.0250	9.8	0.035	11.40	0.1465
	400	10240	36	0.157	0.0187	10.8	0.035	12.56	0.1465
	500	10120	34	0.136	0.0144	12.0	0.035	13.96	0.1465
ALLOY 25% Na + 75% K $t_m = -11°C$ $t_b = 784°C$	100	847	88	0.607	0.0248	20.5	0.273	23.84	1.1430
	200	822	95	0.452	0.0171	20.0	0.256	23.26	1.0718
	300	799	99	0.366	0.0134	19.5	0.248	22.68	1.0383
	400	775	103	0.308	0.0108	19.0	0.239	22.10	1.0006
	500	751	107	0.267	0.0090	18.5	0.231	21.52	0.9672

t_m, melting point; t_b, boiling point.

PROPERTY VALUES OF GASES AT ONE ATMOSPHERIC PRESSURE

Temperature °C	Density ρ kg/m³	Kinematic viscosity $\nu \times 10^6$ m²/s	Prandtl Number Pr	Thermal Diffusivity $\alpha \times 10^3$ m²/hr	Specific Heat C_p kcal/kg°C	Thermal Conductivity $k \times 10^3$ kcal/m-hr°C	Coefficient of viscosity $\mu \times 10^6$ kgf s/m²	Specific Heat C_p kJ/kg K	Thermal Conductivity $k \times 10^3$ W/mK	Coefficient of viscosity $\mu \times 10^6$ Ns/m² or kg/ms
DRY AIR										
-50	1.584	9.23	0.728	45.7	0.242	17.5	1.49	1.013	20.35	14.61
-40	1.515	10.04	0.728	49.6	0.242	18.2	1.55	1.013	21.17	15.20
-30	1.453	10.80	0.723	53.7	0.242	18.9	1.60	1.013	21.98	15.69
-20	1.395	11.61	0.716	58.3	0.241	19.6	1.65	1.009	22.79	16.18
-10	1.342	12.43	0.712	52.8	0.241	20.3	1.70	1.009	23.61	16.67
0	1.293	13.28	0.707	67.7	0.240	21.0	1.75	1.005	24.42	17.16
10	1.247	14.16	0.705	72.2	0.240	21.6	1.80	1.005	25.12	17.65
20	1.205	15.06	0.703	77.1	0.240	22.3	1.85	1.005	25.93	18.14
30	1.165	16.00	0.701	82.3	0.240	23.0	1.90	1.005	26.75	18.63
40	1.128	16.96	0.699	87.5	0.240	23.7	1.95	1.005	27.56	19.12
50	1.093	17.95	0.698	92.6	0.240	24.3	2.00	1.005	28.26	19.61
60	1.060	18.97	0.696	97.9	0.240	24.9	2.05	1.005	28.96	20.10
70	1.029	20.02	0.694	102.8	0.241	25.5	2.10	1.009	29.66	20.59
80	1.000	21.09	0.692	108.7	0.241	26.2	2.15	1.009	30.47	21.08
90	0.972	22.10	0.690	114.8	0.241	26.9	2.19	1.009	31.28	21.48
100	0.946	23.13	0.688	121.1	0.241	27.6	2.23	1.009	32.10	21.87
120	0.898	25.45	0.686	132.6	0.241	28.7	2.33	1.009	33.38	22.85
140	0.854	27.80	0.684	145.2	0.242	30.0	2.42	1.013	34.89	23.73
160	0.815	30.09	0.682	158.0	0.243	31.3	2.50	1.017	36.40	24.52
180	0.779	32.49	0.681	171.0	0.244	32.5	2.58	1.022	37.80	25.30
200	0.746	34.85	0.680	184.9	0.245	33.8	2.65	1.026	39.31	25.99
250	0.674	40.61	0.677	210.6	0.248	36.7	2.79	1.038	42.68	27.36
300	0.615	48.20	0.674	257.6	0.250	39.6	3.03	1.047	46.05	29.71
350	0.566	55.46	0.676	294.7	0.253	42.2	3.20	1.059	49.08	31.38
400	0.524	63.09	0.678	335.2	0.255	44.8	3.37	1.067	52.10	33.05
500	0.456	79.38	0.687	415.1	0.261	49.4	3.69	1.093	57.45	36.19
600	0.404	96.89	0.699	499.0	0.266	53.5	3.99	1.114	62.22	39.13
700	0.362	115.40	0.706	588.2	0.271	57.5	4.26	1.135	66.87	41.78
800	0.329	134.80	0.713	682.0	0.276	61.7	4.52	1.156	71.76	44.33

PROPERTY VALUES OF GASES AT ONE ATMOSPHERIC PRESSURE (Cont.)

Temperature °C	Density ρ kg/m³	Kinematic viscosity ν x 10⁶ m²/s	Prandtl Number Pr	Thermal Diffusivity α x 10³ m²/hr	Specific Heat Cp kcal/kg°C	Thermal Conductivity k x 10³ kcal/m-hr°C	Coefficient of viscosity μ x 10⁶ kgf s/m²	Specific Heat Cp kJ/kg K	Thermal Conductivity k x 10³ W/mK	Coefficient of viscosity μ x 10⁶ Ns/m² or kg/m/s
900	0.301	155.10	0.717	778.4	0.280	65.6	4.76	1.172	76.29	46.68
1000	0.277	178.00	0.719	888.0	0.283	69.4	5.00	1.185	80.71	49.03
1100	0.257	199.30	0.722	994.5	0.286	73.1	5.22	1.197	85.02	51.19
1200	0.239	223.70	0.724	1139.4	0.289	78.7	5.45	1.210	91.53	53.45
NITROGEN										
0	1.250	13.3	0.705	68.9	0.246	20.9	1.70	1.030	24.31	16.67
100	0.916	22.5	0.678	116	0.247	27.1	2.11	1.034	31.52	20.69
200	0.723	33.6	0.656	183	0.249	33.1	2.47	1.043	38.50	24.22
300	0.597	46.4	0.652	255	0.253	38.6	2.82	1.059	44.89	27.65
400	0.508	60.9	0.659	333	0.258	43.6	3.15	1.080	50.71	30.89
500	0.442	76.9	0.672	411	0.264	48.0	3.46	1.105	55.82	33.93
600	0.392	94.3	0.689	491	0.270	51.9	3.76	1.130	60.36	36.87
700	0.352	113.0	0.710	570	0.275	55.2	4.04	1.151	64.20	39.62
800	0.318	133.0	0.734	654	0.279	58.0	4.31	1.168	67.45	42.27
900	0.291	154.0	0.762	731	0.284	60.3	4.59	1.189	70.13	45.01
1000	0.268	177.0	0.795	802	0.287	62.2	4.84	1.202	72.34	47.46
OXYGEN										
0	1.429	13.6	0.720	68	0.218	21.2	1.98	0.913	24.66	19.42
100	1.050	23.1	0.686	121	0.223	28.3	2.46	0.934	32.91	24.12
200	0.826	34.6	0.674	156	0.230	35.0	2.91	0.963	40.71	28.54
300	0.682	47.8	0.673	254	0.238	41.3	3.31	0.997	48.03	32.46
400	0.580	62.8	0.675	333	0.244	47.3	3.70	1.022	55.01	36.28
500	0.504	79.6	0.682	420	0.250	52.9	4.08	1.047	61.52	40.01
600	0.447	97.8	0.689	508	0.255	58.0	4.44	1.068	67.45	43.54
700	0.402	117.0	0.700	600	0.259	62.6	4.79	1.084	72.80	46.97
800	0.363	138.0	0.710	700	0.263	66.8	5.12	1.101	77.69	50.21
900	0.333	161.0	0.725	797	0.266	70.5	5.45	1.114	81.99	53.45
1000	0.306	184.0	0.738	900	0.268	73.8	5.76	1.122	85.83	56.49

PROPERTY VALUES OF GASES AT ONE ATMOSPHERIC PRESSURE (Cont.)

Temperature °C	Density ρ kg/m³	Kinematic viscosity ν x 10⁶ m²/s	Prandtl Number Pr	Thermal Diffusivity α x 10³ m²/hr	Specific Heat C_p kcal/kg°C	Thermal Conductivity k x 10³ kcal/m·hr°C	Coefficient of viscosity μ x 10⁶ kgf s/m²	Specific Heat C_p kJ/kg K	Thermal Conductivity k x 10³ W/mK	Coefficient of viscosity μ x 10⁶ Ns/m² or kg/m/ms
CARBON MONOXIDE										
0	1.250	13.3	0.740	64.6	0.248	20.0	1.69	1.038	23.26	16.57
100	0.916	22.6	0.718	113	0.249	25.9	2.11	1.043	30.12	20.69
200	0.723	33.9	0.708	179	0.253	31.4	2.49	1.059	36.52	24.42
300	0.596	47.0	0.709	238	0.258	36.6	2.85	1.080	42.57	27.95
400	0.508	61.8	0.711	311	0.264	41.7	3.18	1.105	48.50	31.19
500	0.442	78.0	0.720	389	0.270	46.5	3.51	1.130	54.08	34.42
600	0.392	96.0	0.727	474	0.276	51.3	3.81	1.156	59.66	37.36
700	0.351	115.0	0.733	566	0.282	55.9	4.12	1.181	65.01	40.40
800	0.317	135.0	0.739	667	0.286	60.3	4.41	1.197	70.13	43.25
900	0.291	157.0	0.740	768	0.290	64.9	4.69	1.214	75.48	45.99
1000	0.268	180.0	0.744	881	0.294	69.3	4.97	1.231	80.60	48.74
CARBON DIOXIDE										
0	1.977	7.09	0.780	32.8	0.195	12.6	1.43	0.816	14.65	14.02
100	1.447	12.60	0.733	62.1	0.218	19.6	1.86	0.913	22.79	18.24
200	1.143	19.20	0.715	98.3	0.237	26.6	2.28	0.992	30.94	22.36
300	0.944	27.30	0.712	141	0.252	33.6	2.69	1.055	39.08	26.38
400	0.802	36.70	0.709	191	0.265	40.6	3.08	1.109	47.22	30.20
500	0.698	47.20	0.713	246	0.276	47.2	3.46	1.156	54.89	33.93
600	0.618	58.30	0.723	308	0.285	53.4	3.84	1.193	62.10	37.66
700	0.555	71.40	0.730	366	0.292	59.2	4.19	1.223	68.85	41.09
800	0.502	85.30	0.741	432	0.298	64.6	4.55	1.248	75.13	44.62
900	0.460	100.00	0.757	499	0.304	69.6	4.91	1.273	80.94	48.15
1000	0.423	116.00	0.770	569	0.308	74.2	5.25	1.290	86.29	51.48
SULPHUR DIOXIDE										
0	2.926	4.14	0.874	17.0	0.145	7.2	1.23	0.607	8.37	12.06
100	2.140	7.51	0.863	31.4	0.158	10.6	1.64	0.662	12.33	16.08
200	1.690	11.80	0.856	44.8	0.170	14.3	2.04	0.712	16.63	20.01

PROPERTY VALUES OF GASES AT ONE ATMOSHPERIC PRESSURE (Cont.)

Tempe-rature °C	Density ρ kg/m³	Kinematic viscosity ν x 10⁶ m²/s	Prandtl Number Pr	Thermal Diffusivity α x 10³ m²/hr	Specific Heat C_p kcal/kg°C	Thermal Conductivity k x 10³ kcal/m-hr°C	Coefficient of viscosity μ x 10⁶ kgf s/m²	Specific Heat C_p kJ/kg K	Thermal Conductivity k x 10³ W/mK	Coefficient of viscosity μ x 10⁶ Ns/m² or kg/ms
300	1.395	17.10	0.848	72.5	0.180	18.3	2.43	0.754	21.28	23.83
400	1.185	23.30	0.834	100	0.187	22.2	2.81	0.783	25.82	27.56
500	1.033	30.40	0.822	132	0.193	26.4	3.19	0.808	30.70	31.28
600	0.916	38.30	0.806	170	0.197	30.8	3.57	0.825	35.80	35.01
700	0.892	46.80	0.788	215	0.200	35.3	3.94	0.837	41.05	38.64
800	0.743	56.50	0.774	264	0.203	39.8	4.30	0.850	46.29	42.17
900	0.681	66.80	0.755	320	0.205	44.6	4.66	0.858	51.87	45.70
1000	0.626	78.30	0.740	382	0.207	49.5	5.02	0.867	57.57	49.23
FLUE GASES										
0	1.295	12.20	0.720	60.8	0.249	19.6	1.609	1.043	22.79	15.78
100	0.950	21.54	0.690	111	0.255	26.9	2.079	1.068	31.28	20.38
200	0.748	32.80	0.670	176	0.262	34.5	2.497	1.097	40.12	24.49
300	0.617	45.81	0.650	252	0.268	41.6	2.878	1.122	48.38	28.22
400	0.525	60.38	0.640	339	0.275	49.0	3.230	1.151	56.99	31.68
500	0.457	76.30	0.630	436	0.283	56.4	3.553	1.185	65.59	34.84
600	0.405	93.61	0.620	543	0.290	63.8	3.860	1.214	74.19	37.85
700	0.363	112.10	0.610	662	0.296	71.1	4.148	1.239	82.69	40.68
800	0.329	131.80	0.600	791	0.302	78.7	4.422	1.264	91.53	43.37
900	0.301	152.50	0.590	929	0.308	86.1	4.680	1.290	100.13	45.90
1000	0.275	174.30	0.580	1092	0.312	93.7	4.930	1.306	108.97	48.35
1100	0.257	197.10	0.570	1244	0.316	101	5.169	1.323	117.46	50.69
1200	0.240	221.00	0.560	1413	0.320	109	5.402	1.340	126.77	52.98
ARGON										
0	1.784	11.8	0.663	64.1	0.124	14.2	2.15	0.519	16.51	21.08
100	1.305	20.6	0.661	112	0.124	18.2	2.75	0.519	21.17	26.97
200	1.030	31.2	0.653	172	0.124	22.0	3.28	0.519	25.59	32.17
300	0.850	43.4	0.640	244	0.124	25.7	3.76	0.519	29.89	36.87
400	0.724	56.7	0.628	326	0.124	29.2	4.19	0.519	33.96	41.09
500	0.627	72.0	0.619	420	0.124	32.6	4.61	0.519	37.91	45.21
600	0.558	87.0	0.604	519	0.124	33.9	4.95	0.519	39.43	48.54

PROPERTY VALUES OF GASES AT ONE ATMOSPHERIC PRESSURE (Cont.)

Temperature °C	Density ρ kg/m³	Kinematic viscosity $\nu \times 10^6$ m²/s	Prandtl Number Pr	Thermal Diffusivity $\alpha \times 10^3$ m²/hr	Specific Heat C_p kcal/kg°C	Thermal Conductivity $k \times 10^3$ kcal/m-hr°C	Coefficient of viscosity $\mu \times 10^6$ kgf s/m²	Specific Heat C_p kJ/kg K	Thermal Conductivity $k \times 10^3$ W/mK	Coefficient of viscosity $\mu \times 10^6$ Ns/m² or kg/ms
HELIUM										
0	0.178	105	0.684	552	1.243	123	1.91	5.204	143.04	18.73
100	0.130	176	0.667	948	1.243	154	2.34	5.204	179.10	22.95
200	0.103	270	0.660	1430	1.243	183	2.75	5.204	212.83	26.97
300	0.085	362	0.656	1990	1.243	210	3.14	5.204	244.23	30.79
400	0.072	474	0.648	2630	1.243	237	3.50	5.204	275.63	34.32
500	0.063	611	0.642	3360	1.243	262	3.83	5.204	304.71	37.56
600	0.056	723	0.631	4120	1.243	286	4.11	5.204	332.62	40.31
HYDROGEN										
0	0.0899	93	0.688	486	3.3604	148	0.852	14.069	172.12	8.36
100	0.0657	157	0.677	834	3.4509	189	1.05	14.482	219.81	10.30
200	0.0519	233	0.666	1260	3.4643	227	1.23	14.504	264.00	12.06
300	0.0428	323	0.655	1780	3.4712	264	1.41	14.533	307.03	13.83
400	0.0364	423	0.644	2360	3.4826	299	1.57	14.581	347.74	15.40
500	0.0317	534	0.640	3000	3.5020	333	1.72	14.662	387.28	16.77
600	0.0281	656	0.635	3700	3.5298	367	1.87	14.779	426.82	18.34
700	0.0252	785	0.637	4430	3.5660	398	2.01	14.930	462.87	19.71
800	0.0228	924	0.638	5230	3.6101	430	2.15	15.115	500.09	21.08
900	0.0209	1070	0.640	6030	3.6572	461	2.28	15.312	536.14	22.36
1000	0.0192	1230	0.644	6880	3.7063	491	2.42	15.518	571.03	23.73
AMMONIA										
0	0.771	12.2	0.908	48.1	0.488	18.1	0.954	2.043	21.05	9.36
100	0.564	23.2	0.852	97.8	0.530	29.2	1.33	2.219	33.96	13.04
200	0.445	38.0	0.818	165	0.573	42.0	1.70	2.399	48.85	16.67
300	0.368	56.4	0.812	248	0.617	56.3	2.10	2.583	65.48	20.59
400	0.313	78.7	0.796	351	0.656	72.2	2.48	2.747	83.97	24.32
500	0.272	105.0	0.793	470	0.697	89.1	2.87	2.918	103.62	28.15
600	0.241	134.0	0.792	606	0.736	107	3.26	3.082	124.44	31.97

PROPERTY VALUES OF GASES AT ONE ATMOSPHERIC PRESSURE (Cont.)

Temperature °C	Density ρ kg/m³	Kinematic viscosity $\nu \times 10^6$ m²/s	Prandtl Number Pr	Thermal Diffusivity $\alpha \times 10^3$ m²/hr	Specific Heat C_p kcal/kg°C	Thermal Conductivity $k \times 10^3$ kcal/m-hr°C	Coefficient of viscosity $\mu \times 10^6$ kgf s/m²	Specific Heat C_p kJ/kg K	Thermal Conductivity $k \times 10^3$ W/mK	Coefficient of viscosity $\mu \times 10^6$ Ns/m² or kg/ms
700	0.217	168.0	0.791	758	0.775	127	3.67	3.245	147.70	35.99
800	0.196	205.0	0.793	927	0.813	147	4.06	3.404	170.96	39.82
900	0.179	247.0	0.798	1110	0.849	169	4.50	3.555	196.55	44.13
1000	0.165	291.0	0.800	1310	0.886	191	4.88	3.710	222.13	47.86
TOLUENE - C_7H_8										
0	-	-	0.748	-	0.246	11.1	0.674	1.030	12.91	6.61
100	-	-	-	-	0.337	-	0.903	1.411	-	8.86
200	2.38	4.65	-	-	0.418	-	1.123	1.750	-	11.01
300	1.96	6.75	-	-	0.489	-	1.35	2.047	-	13.24
400	1.667	9.23	-	-	0.548	-	1.57	2.294	-	15.40
500	1.45	12.00	-	-	0.598	-	1.78	2.504	-	17.46
600	1.28	15.30	-	-	0.638	-	2.00	2.671	-	19.61
CARBON TETRA CHLORIDE (REFRIGERANT-10) - CCl_4										
0	-	-	0.802	-	0.1242	5.15	0.942	0.520	5.99	9.24
100	5.020	2.45	0.828	10.6	0.1404	7.51	1.255	0.588	8.73	12.31
200	3.970	3.86	0.816	17.0	0.1482	10.0	1.56	0.621	11.63	15.30
300	3.275	5.59	0.796	25.2	0.1550	12.6	1.86	0.641	14.65	18.24
400	2.790	7.64	0.776	35.1	0.1562	15.3	2.16	0.654	17.79	21.18
500	2.420	9.96	0.758	47.2	0.1593	18.2	2.45	0.667	21.17	24.03
600	2.150	12.60	0.741	60.8	0.1614	21.1	2.74	0.676	24.54	26.87
ACETONE - C_3H_6O										
0	-	-	0.886	-	0.300	8.36	0.700	1.256	9.72	6.86
100	1.87	5.07	0.840	21.8	0.367	14.9	0.960	1.537	17.33	9.41
200	1.47	8.22	0.806	36.8	0.427	23.1	1.230	1.788	26.87	12.06
300	1.22	12.10	0.774	56.4	0.483	33.2	1.500	2.022	38.51	14.71
400	1.03	16.90	0.743	81.5	0.534	44.8	1.770	2.236	52.10	17.36
500	0.901	22.30	0.720	111.0	0.580	58.0	2.040	2.428	67.45	20.01
600	0.799	28.30	0.695	147.0	0.618	72.8	2.320	2.587	84.67	22.75

PROPERTY VALUES OF GASES AT ONE ATMOSPHERIC PRESSURE (Cont.)

Temperature °C	Density ρ kg/m³	Kinematic viscosity $\nu \times 10^6$ m²/s	Prandtl Number Pr	Thermal Diffusivity $\alpha \times 10^3$ m²/hr	Specific Heat C_p kcal/kg°C	Thermal Conductivity $k \times 10^3$ kcal/m-hr°C	Coefficient of viscosity $\mu \times 10^6$ kgf s/m²	Specific Heat C_p kJ/kg K	Thermal Conductivity $k \times 10^3$ W/mK	Coefficient of viscosity $\mu \times 10^6$ Ns/m² or kg/ms
BENZENE - C_6H_6										
0	-	-	0.716	-	0.2253	7.93	0.712	0.943	9.22	6.98
100	2.55	3.74	0.554	18.4	0.3189	14.9	0.735	1.335	17.33	7.21
200	2.01	5.99	0.719	30.1	0.4003	24.2	1.233	1.676	28.14	12.09
300	1.66	8.80	0.688	46.2	0.4673	35.8	1.493	1.957	41.64	14.64
400	1.41	12.10	0.652	67.0	0.5213	49.5	1.754	2.183	57.57	17.20
500	1.23	15.90	0.614	94.4	0.5659	65.7	2.015	2.369	76.41	19.76
600	1.09	20.40	0.585	126.0	0.6029	82.8	2.275	2.524	96.30	22.31
STEAM										
100	0.598	20.0	1.08	69.2	0.510	20.4	1.22	2.135	23.73	11.96
200	0.464	30.6	0.94	132	0.472	28.8	1.62	1.976	33.49	15.89
300	0.384	44.3	0.91	206	0.481	38.0	2.04	2.014	44.19	20.01
400	0.326	60.5	0.90	298	0.495	48.1	2.48	2.073	55.94	24.32
500	0.284	78.8	0.90	406	0.510	58.8	2.92	2.135	68.38	28.64
600	0.252	99.8	0.89	531	0.527	70.3	3.38	2.206	81.76	33.15
700	0.226	122	0.90	670	0.543	82.2	3.86	2.273	95.60	37.85
800	0.204	147	0.91	829	0.560	94.8	4.34	2.345	110.25	42.56
900	0.187	174	0.92	993	0.577	107	4.84	2.416	124.44	47.46
1000	0.172	204	0.92	1190	0.593	121	5.34	2.483	140.72	52.37

SATURATED STEAM

Saturation temperature °C	Density ρ kg/m³	Kinematic viscosity ν x 10⁶ m²/s	Prandtl Number Pr	Thermal Diffusivity α x 10⁶ m²/hr	Coefficient of viscosity μ x 10⁶ kgf s/m²	Saturation Pressure p kgf/cm²	Specific Heat C_p kcal/kg°C	Thermal Conductivity k x 10³ kcal/mhr°C	Coefficient of viscosity μ x 10⁶ Ns/m²or kg/ms	Saturation Pressure p bar	Specific Heat C_p kJ/kg K	Thermal Conductivity k x 10³ W/mK
100	0.598	20.020	1.08	66.9	1.22	1.03	0.510	20.4	11.96	1.013	2.135	23.73
110	0.826	15.070	1.09	49.8	1.27	1.46	0.520	21.4	12.45	1.433	2.177	24.89
120	1.12	11.460	1.09	37.8	1.31	2.02	0.527	22.3	12.85	1.985	2.206	25.93
130	1.50	8.850	1.11	28.7	1.35	2.75	0.539	23.1	13.24	2.701	2.257	26.87
140	1.97	6.890	1.12	22.07	1.38	3.69	0.553	24.1	13.53	3.614	2.315	28.03
150	2.55	5.470	1.15	17.02	1.42	4.85	0.572	24.8	13.93	4.760	2.395	28.84
160	3.26	4.390	1.18	13.40	1.46	6.30	0.592	25.9	14.32	6.181	2.479	30.12
170	4.12	3.570	1.21	10.58	1.50	8.08	0.617	26.9	14.71	7.920	2.583	31.28
180	5.16	2.930	1.25	8.42	1.54	10.23	0.647	28.1	15.10	10.027	2.709	32.68
190	6.40	2.440	1.30	6.74	1.59	12.80	0.682	29.4	15.59	12.551	2.855	34.19
200	7.86	2.030	1.34	5.37	1.63	15.86	0.710	30.5	15.98	15.549	2.973	35.47
210	9.58	1.710	1.41	4.37	1.67	19.46	0.740	32.0	16.37	19.077	3.098	37.22
220	11.6	1.450	1.47	3.54	1.72	23.66	0.814	33.5	16.87	23.198	3.408	38.96
230	14.0	1.240	1.54	2.90	1.77	28.53	0.868	35.2	17.36	27.976	3.634	40.94
240	16.8	1.060	1.61	2.37	1.81	34.14	0.927	36.9	17.50	33.478	3.881	42.91
250	20.0	0.913	1.68	1.96	1.86	40.56	0.993	38.8	18.24	39.776	4.158	45.12
260	23.7	0.794	1.75	1.63	1.92	47.87	1.067	41.3	18.83	46.943	4.467	48.03
270	28.1	0.688	1.82	1.36	1.97	56.14	1.150	43.9	19.32	55.058	4.815	51.06
280	33.2	0.600	1.90	1.14	2.03	65.46	1.250	47.2	19.91	64.202	5.234	54.89
290	39.2	0.526	2.01	0.94	2.10	75.92	1.360	50.1	20.59	74.461	5.694	58.27

SATURATED STEAM (Cont.)

Saturation temperature °C	Density ρ kg/m³	Kinematic viscosity ν x 10⁶ m²/s	Prandtl Number Pr	Thermal Diffusivity α x 10⁶ m²/hr	Coefficient of viscosity μ x 10⁶ kgf s/m²	Saturation Pressure p kgf/cm²	Specific Heat C_p kcal/kg°C	Thermal Conductivity k x 10³ kcal/mhr°C	Coefficient of viscosity μ x 10⁶ Ns/m² or kg/ms	Saturation Pressure p bar	Specific Heat C_p kJ/kg K	Thermal Conductivity k x 10³ W/mK
300	46.2	0.461	2.13	0.78	2.17	87.61	1.500	53.9	21.28	85.927	6.280	62.69
310	54.6	0.403	2.29	0.63	2.24	100.64	1.700	58.8	21.97	98.70	7.118	68.38
320	64.7	0.353	2.50	0.51	2.33	115.12	1.960	64.6	22.85	112.89	8.206	75.13
330	77.0	0.310	2.86	0.39	2.44	131.18	2.360	71.0	23.93	128.63	9.881	82.57
340	92.8	0.272	3.35	0.29	2.57	148.96	2.950	80.0	25.20	146.05	12.351	93.04
350	114	0.234	4.03	0.21	2.71	168.63	3.880	92.0	26.58	165.35	16.245	107.00
360	144	0.202	5.23	0.14	2.97	190.42	5.500	110	29.13	186.75	23.027	127.93
370	202	0.166	11.10	0.05	3.44	214.68	13.500	147	33.73	210.54	56.522	170.96

1 bar = 10⁵ N/m²

THERMAL CONDUCTIVITY OF AIR AT VARIOUS PRESSURES AND TEMPERATURES

Unit : W/m K

Temperature °C / Pressure bar	20	100	180
1	0.0258	0.0307	0.0362
100	0.0280	0.0309	0.0367
200	0.0333	0.0372	0.0409
300	0.0385	0.0434	0.0450
400	0.0510	0.0475	0.0485

Unit : kcal/m-hr °C

Temperature °C / Pressure kgf/cm²	20	100	180
1	0.0221	0.0263	0.0311
100	0.0239	0.0265	0.0314
200	0.0328	0.0323	0.0351
300	0.0390	0.0371	0.0384
400	0.0435	0.0404	0.0416

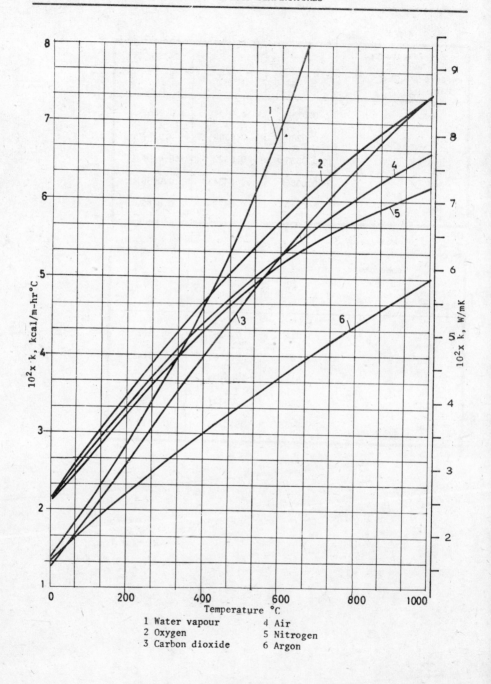

1 Water vapour   4 Air
2 Oxygen   5 Nitrogen
3 Carbon dioxide 6 Argon

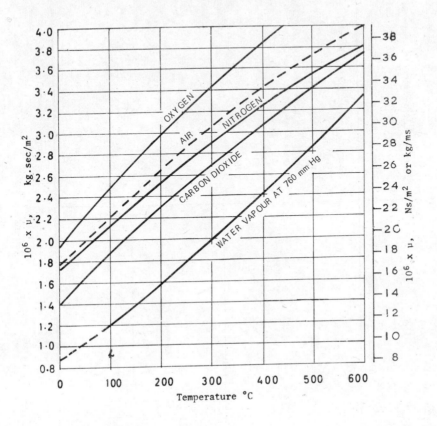

ONE DIMENSIONAL STEADY STATE HEAT CONDUCTION

HEAT CONDUCTION — THERMAL RESISTANCE OF BODIES

Heat Flow, $Q = \dfrac{\Delta T \text{ overall}}{R}$

ΔT, overall difference in temperature, K

R, thermal resistance, hr K/k cal or K/W

Q, heat flow, k cal/h or W

k, thermal conductivity, k cal/h. m. k or W/mK

A, area of heat flow, m^2

h, convective heat transfer coefficient, k cal/h. m^2. K or W/m^2 K

$\Delta T = T_a - T_b$

Shape	Thermal Resistance, R
Plane wall	$\dfrac{L}{kA}$
Composite plane wall	$\dfrac{1}{A}\left[\dfrac{1}{h_a} + \dfrac{L_1}{k_1} + \dfrac{L_2}{k_2} + \dfrac{L_3}{k_3} + \dfrac{1}{h_b}\right]$

34

ONE DIMENSIONAL STEADY STATE HEAT CONDUCTION (Cont.)

Shape	Thermal Resistance, R	
Composite cylinder	$$\frac{1}{2\pi\ell}\left[\frac{1}{h_a r_1} + \frac{1}{k_1}\ln\left(\frac{r_2}{r_1}\right) + \frac{1}{k_2}\ln\left(\frac{r_3}{r_2}\right) + \frac{1}{k_3}\ln\left(\frac{r_4}{r_3}\right) + \frac{1}{h_b r_4}\right]$$	r, radius, m ℓ, length, m
Hollow sphere	$$\frac{1}{4\pi}\left[\frac{1}{h_a r_1^2} + \frac{1}{k}\left(\frac{1}{r_1} - \frac{1}{r_2}\right) + \frac{1}{h_b r_2^2}\right]$$	

35

ONE DIMENSIONAL STEADY STATE HEAT CONDUCTION (Cont.)

Shape	Thermal Resistance, R
Pipe in square section	$$\approx \frac{1}{2\pi\ell} \left\{ \frac{1}{h_a r} + \frac{1}{k} \ln\left(\frac{1.08\, a}{2r}\right) + \frac{\pi}{2 h_b a} \right\}$$
Pipe with eccentric lagging of length ℓ	$$\frac{1}{2\pi k\ell} \ln \frac{\sqrt{[(r_2 + r_1)^2 - e^2]} + \sqrt{[(r_2 - r_1)^2 - e^2]}}{\sqrt{[(r_2 + r_1)^2 - e^2]} - \sqrt{[(r_2 - r_1)^2 - e^2]}}$$

STEADY STATE CONDUCTION WITH HEAT GENERATION : TEMPERATURE DISTRIBUTION

Shape	Temperature Distribution	
Slab	$$\dfrac{T_x - T_o}{T_w - T_o} = (X/L)^2$$ $$T_x = T_o - \dfrac{\dot{q}}{2k}(X^2)$$	$\dot{q}$, heat generated, kcal/h· m^3 or W/m^3 T_o, temp at mid plane, K T_x, temp at distance X from mid plane, K T_w, wall temp, K L, half thickness, m r, R, radii, m T_r, temp at any radius, K
Solid cylinder	$$\dfrac{T_r - T_w}{T_o - T_w} = 1 - (r/R)^2$$ $$T_r = T_w + \dfrac{\dot{q}}{4k}(R^2 - r^2)$$	

37

FINS OR EXTENDED SURFACES

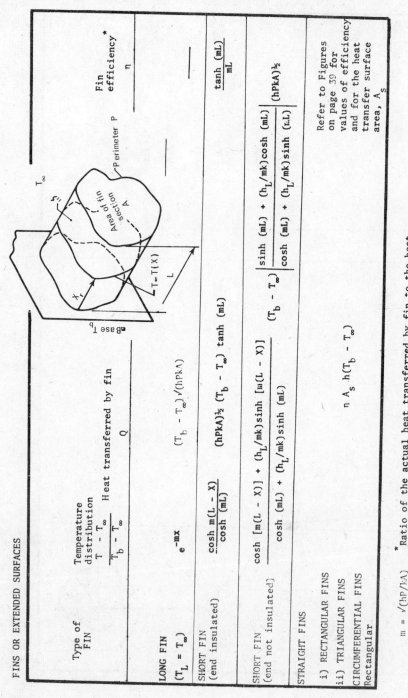

Type of FIN	Temperature distribution $\dfrac{T - T_\infty}{T_b - T_\infty}$	Heat transferred by fin Q	Fin efficiency* η
LONG FIN $(T_L = T_\infty)$	e^{-mx}	$(T_b - T_\infty)\sqrt{(hPkA)}$	___
SHORT FIN (end insulated)	$\dfrac{\cosh m(L - X)}{\cosh (mL)}$	$(hPkA)^{\frac12}\,(T_b - T_\infty)\,\tanh (mL)$	$\dfrac{\tanh (mL)}{mL}$
SHORT FIN (end not insulated)	$\dfrac{\cosh\,[m(L - X)] + (h_L/mk)\sinh\,[m(L - X)]}{\cosh\,(mL) + (h_L/mk)\sinh\,(mL)}$	$(T_b - T_\infty)\left[\dfrac{\sinh\,(mL) + (h_L/mk)\cosh\,(mL)}{\cosh\,(mL) + (h_L/mk)\sinh\,(mL)}\right](hPkA)^{\frac12}$	
STRAIGHT FINS i) RECTANGULAR FINS ii) TRIANGULAR FINS CIRCUMFERENTIAL FINS Rectangular		$\eta\,A_s\,h(T_b - T_\infty)$	Refer to Figures on page 39 for values of efficiency and for the heat transfer surface area, A_s

* Ratio of the actual heat transferred by fin to the heat transferable by fin, if the entire fin area were at base temperature.

$m = \sqrt{(hP/kA)}$

38

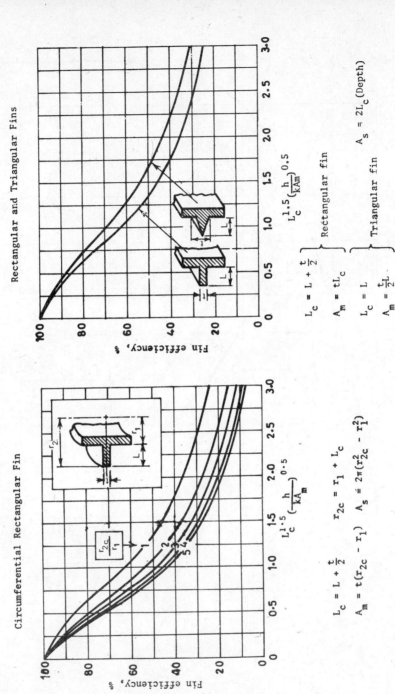

Circumferential Rectangular Fin

Rectangular and Triangular Fins

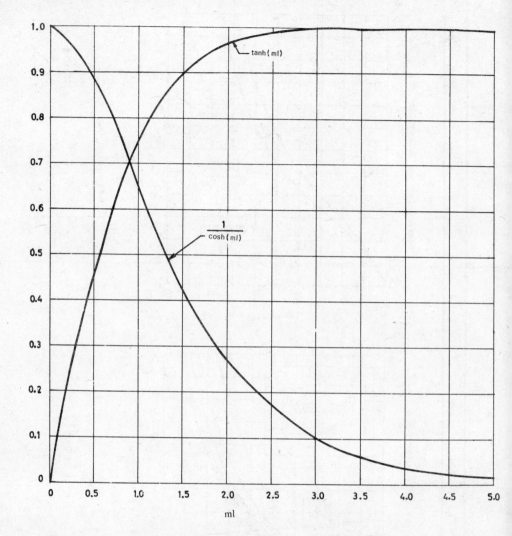

Formula: $Q = kS\Delta T$

Q, heat transferred, kcal/h or W

k, thermal conductivity kcal/h m °C or W/mK

ΔT, overall temperature difference, °C or K

S, SHAPE FACTOR, m

Shape	SHAPE FACTOR, S
Plane wall	$\dfrac{A}{L}$
Hollow cylinder	$\dfrac{2\pi\,L}{\ln\left(\dfrac{r_o}{r_i}\right)}$ for $L \gg r$
Hollow sphere	$\dfrac{4\pi\,r_i r_o}{(r_o - r_i)}$

Shape	SHAPE FACTOR
 Buried cylinder of length L	Case (i) For $r \ll L$ and $D > 3r$ $\dfrac{2\pi L}{\ln \left(\dfrac{2D}{r}\right)}$ Case (ii) For $r \ll L$ $\dfrac{2\pi L}{\cosh^{-1}(D/r)}$
	WALLS; $\dfrac{A}{L}$ EDGES: 0.54 D CORNERS: 0.15 L

42

TWO DIMENSIONAL NODAL EQUATIONS:

	Description	Equation for Residue
	Interior Node	$T_{m+1,n} + T_{m,n+1} + T_{m-1,n} + T_{m,n-1} - 4T_{m,n}$
	Convection Boundary	$\dfrac{h\Delta x}{k} T_\infty + \dfrac{1}{2}[2\, T_{m-1,n} + T_{m,n+1} + T_{m,n-1}] - [\dfrac{h\Delta x}{k} + 2]\, T_{m,n}$
	Exterior Corner	$2\dfrac{h\Delta x}{k} T_\infty + [T_{m-1,n} + T_{m,n-1}] - 2[\dfrac{h\Delta x}{k} + 1]\, T_{m,n}$

TWO DIMENSIONAL NODAL EQUATIONS (Cont.):

Diagram	Description	Equation for Residue
Δx, Δy; $m,n+1$, m,n, $m,n-1$, $m-1,n$, $m+1,n$, h, T_∞	Interior Corner	$\dfrac{2h\Delta x}{k}T_\infty + 2T_{m-1,n} + 2T_{m,n+1} + T_{m+1,n}$ $+ T_{m,n-1} - 2[3 + \dfrac{h\Delta x}{k}]T_{m,n}$
Δx, Δy; $m,n+1$, m,n, $m,n-1$, $m-1,n$	Insulated Boundary	$T_{m,n+1} + T_{m,n-1} + 2T_{m-1,n} - 4T_{m,n}$
Δy, $c\Delta y$, $b\Delta y$, $a\Delta x$; $m,n+1$, m,n, $m,n-1$, $m-1,n$, $m+1,n$; 1, 2, 3	Curved Boundary	$\dfrac{2}{b(b+1)}T_2 + \dfrac{2}{a+1}T_{m+1,n} + \dfrac{2}{b+1}T_{m,n-1}$ $+ \dfrac{2}{a(a+1)}T_1 - 2[\dfrac{1}{a} + \dfrac{1}{b}]T_{m,n}$

TRANSIENT TEMPERATURE HISTORY AND HEAT TRANSFER

Description and state of bodies	Temperature distribution θ/θ_i	Heat transferred Q kcal/h or W	
*BODIES WITH NEGLIGIBLE INTERNAL THERMAL RESISTANCE	$e^{-\left[\frac{hA}{C\rho V}\right]\tau}$ also $e^{-(Bi)(Fo)}$ †	heat transfer rate, $dQ/d\tau$ $hA(T - T_f)$	$\theta = T - T_f;\ \theta_i = T_i - T_f$ $\theta_o = T_o - T_f$ h, heat transfer coefficient A, surface area C, specific heat ρ, density V, volume τ, time; T, temperature at any point and/or time
INFINITE PLATE OR SLAB $y = z = \infty$ x = discrete value	Refer charts on pages 47 to 53	Refer chart on page 46	T_f or T_∞, surrounding fluid temp. T_i, T at $\tau = 0$ T_o, T at mid plane or axis or at centre
INFINITE CYLINDER (length infinite)	Refer charts on pages 47, 54-58	Refer chart on page 46	T_m, mean temperature U_i, initial internal energy $= \rho C V \theta_i$
SPHERE	Refer charts on pages 47, 59-63	Refer chart on page 46	Q = total energy change up to time τ
SEMI INFINITE BODIES $x = y = z = \infty$ and at $x = 0$, (θ/θ_i) = Const at all times	Refer also to chart on page 64 $erf[x/2\sqrt{(\alpha\tau)}]$, tabulated on on page 66	$- 2kA\theta_i\sqrt{(\tau/\pi\alpha)}$	k, thermal conductivity α, thermal diffusivity C (II) solution for infinite cylinder P (X) solution for infinite plate S (X) solution for semi infinite bodies
SEMI INFINITE BODIES with convective boundary conditions	Refer chart on page 64	—	
OTHER SOLIDS OF SPECIAL SHAPES	Use product solutions suggested on page 65	—	

*For the case of thermocouple, it is usual to refer to the response time in terms of a time constant.
This time constant is defined as follows:- $\theta/\theta_i = 1/e$; alternately $\tau = C\rho V/hA$

†Bi, Biot Number; Fo, Fourier Number

Heat transfer variation with time

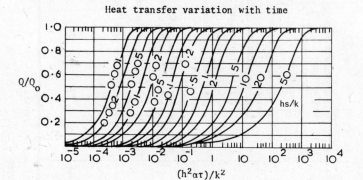

Infinite cylinder of radius, s

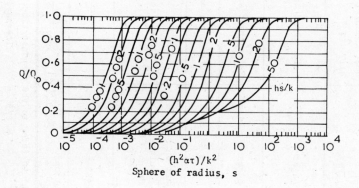

Sphere of radius, s

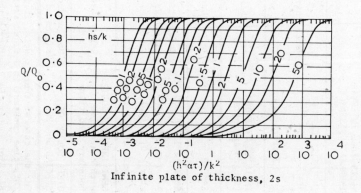

Infinite plate of thickness, 2s

Plate: m = 1
Cylinder: m = 2   value of m
Sphere: m 3
Range of these lines: $\alpha\tau/s^2 > 0.2$, $\frac{h\ s}{k} < 0.01$

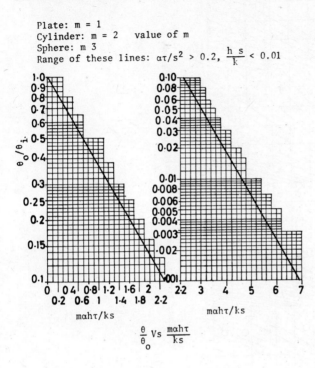

$\frac{\theta}{\theta_o}$ Vs $\frac{m\alpha h\tau}{ks}$

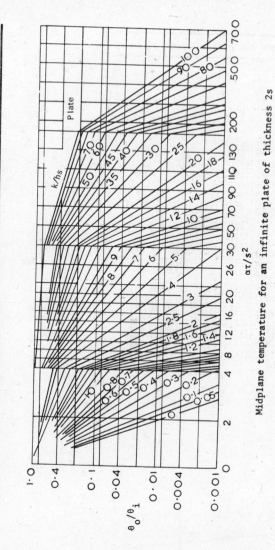

Midplane temperature for an infinite plate of thickness 2s

TRANSIENT TEMPERATURE HISTORY AND HEAT TRANSFER (Cont.)

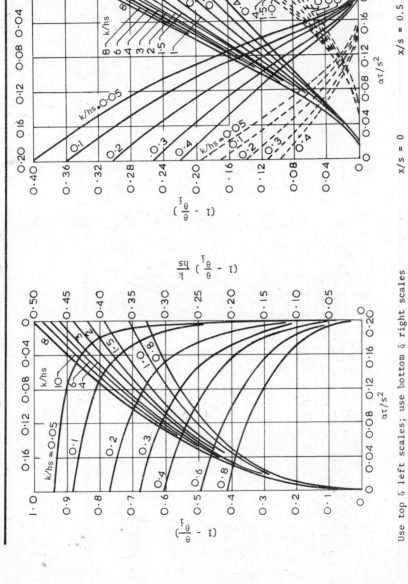

Use top & left scales; use bottom & right scales

$x/s = 0$ $x/s = 0.5$

Use top & left scales; use bottom & right scales

Temperature change in the interior of plane plate after sudden change of environment for $\alpha\tau/s^2 < 0.2$

Use top & left scales; use bottom & right scales

Temperature change on the surface of plane plate $(x/s = 1)$ after sudden change of environment temperature for $\alpha\tau/s^2 < 0.2$

49

TRANSIENT TEMPERATURE HISTORY AND HEAT TRANSFER (Cont.)

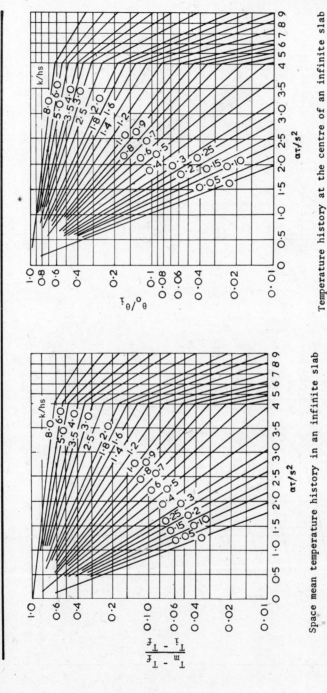

Temperature history at the centre of an infinite slab

Space mean temperature history in an infinite slab

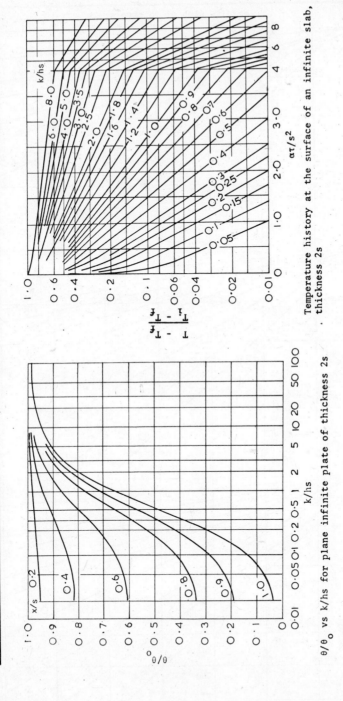

Temperature history at the surface of an infinite slab, thickness 2s

θ/θ_o vs k/hs for plane infinite plate of thickness 2s

Infinite slab of thickness, 2s

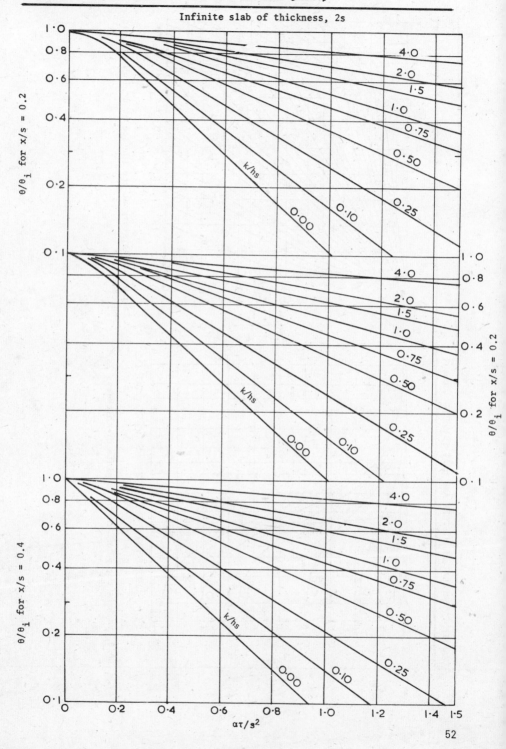

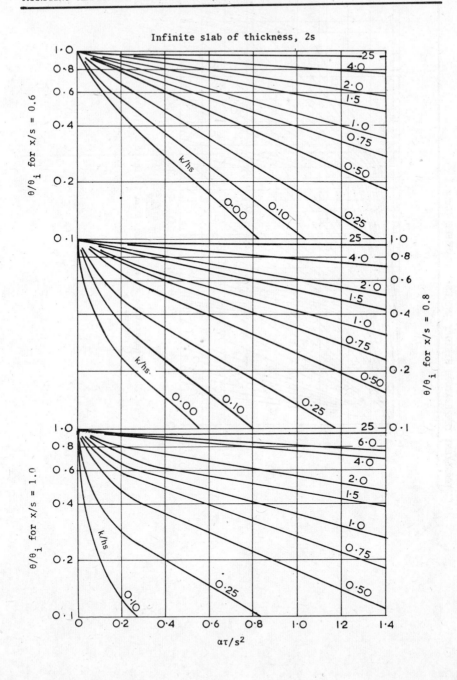

Infinite slab of thickness, 2s

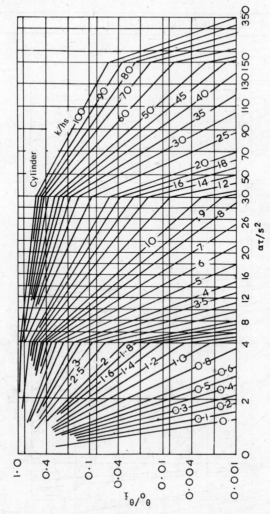

Temperature at the axis of an infinite cylinder of radius, s

54

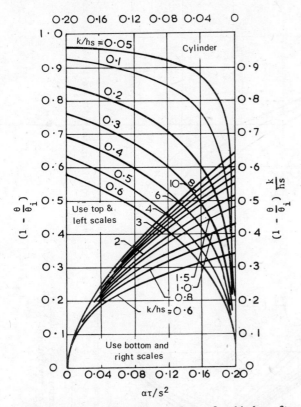

Temperature change on the surface of cylinder after
sudden change of environment temperature for
$\alpha\tau/s^2 < 0.2$ for infinitely long cylinder of radius, s

TRANSIENT TEMPERATURE HISTORY AND HEAT TRANSFER (Cont.)

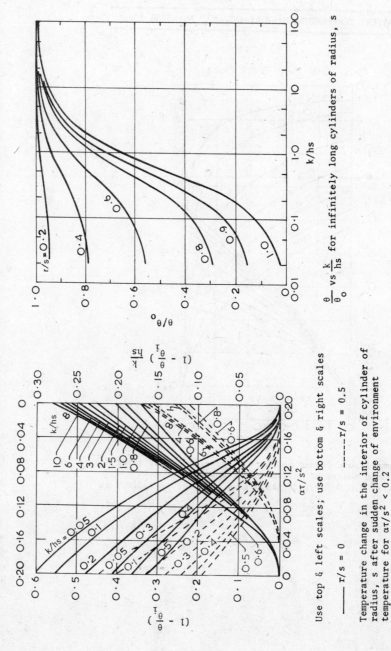

Use top & left scales; use bottom & right scales

——— r/s = 0   ------ r/s = 0.5

$\dfrac{\theta}{\theta_o}$ vs $\dfrac{k}{hs}$ for infinitely long cylinders of radius, s

Temperature change in the interior of cylinder of radius, s after sudden change of environment temperature for $\alpha\tau/s^2 < 0.2$

56

Infinite cylinder of radius, s

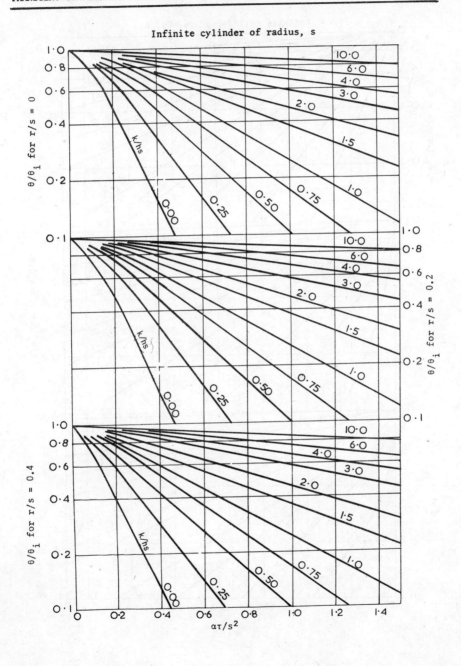

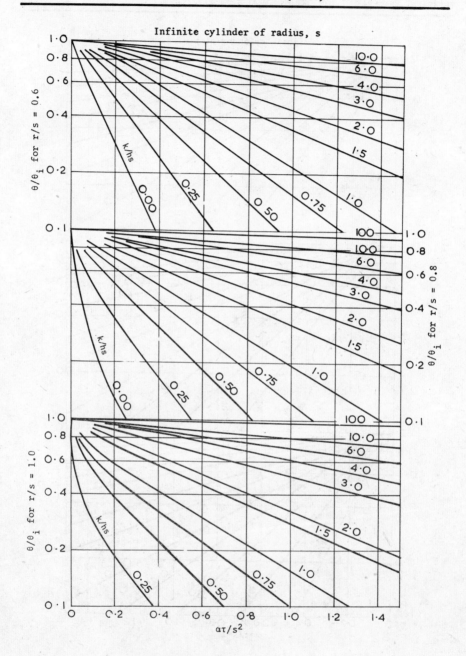

Infinite cylinder of radius, s

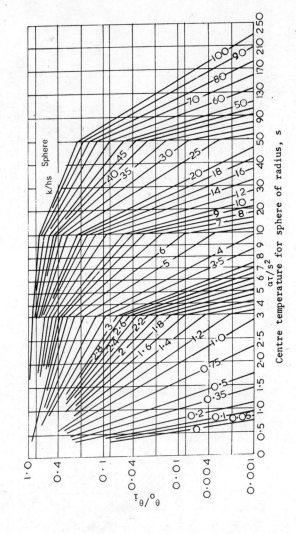

TRANSIENT TEMPERATURE HISTORY AND HEAT TRANSFER (Cont.)

Temperature change at half radius of sphere
after sudden change of environment temperature
for $\alpha\tau/s^2 < 0.2$

Temperature change on the surface and at
the centre of sphere after sudden change
of temperature for $\alpha\tau/s^2 < 0.2$

60

θ/θ_o versus k/hs for sphere, radius s

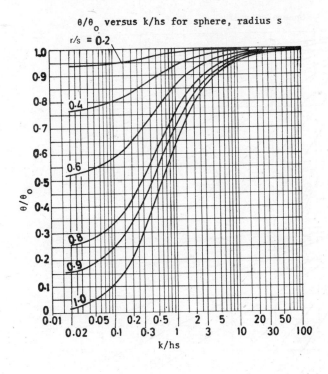

Sphere of radius, s

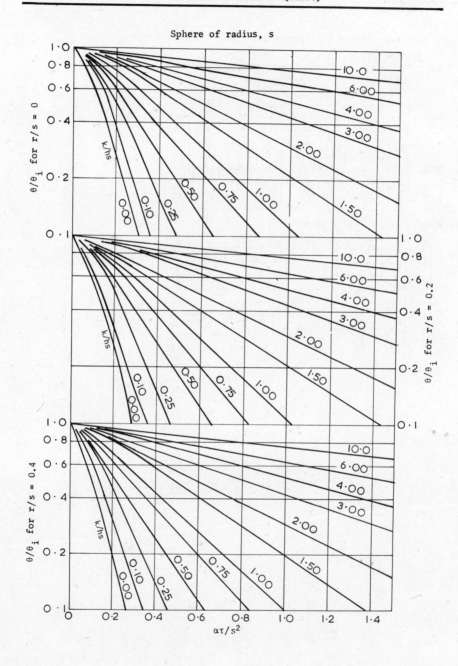

Sphere of radius, s

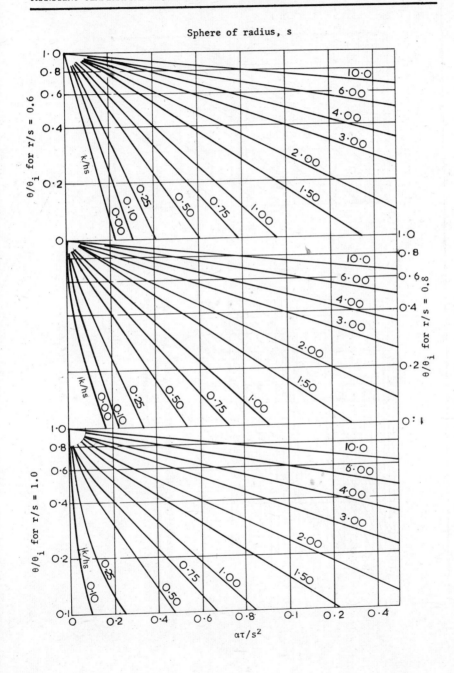

$\alpha\tau/s^2$

$$\frac{T - T_i}{T_\infty - T_i} = 1 - \frac{T - T_\infty}{T_i - T_\infty}$$

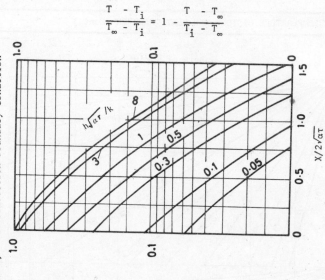

Temperature distribution in the semi infinite body with convection boundary condition

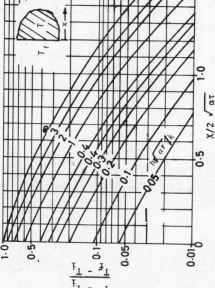

Temperature history in a semi-infinite solid initially at a uniform temperature and suddenly exposed at its surface to a fluid at constant temperature.

64

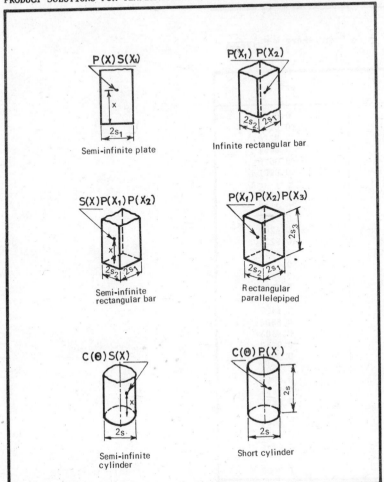

$P(X)S(X_1)$

Semi-infinite plate

$P(X_1) P(X_2)$

Infinite rectangular bar

$S(X)P(X_1)P(X_2)$

Semi-infinite
rectangular bar

$P(X_1)P(X_2)P(X_3)$

Rectangular
parallelepiped

$C(\Theta)S(X)$

Semi-infinite
cylinder

$C(\Theta)P(X)$

Short cylinder

VALUES OF GAUSS' ERROR INTEGRAL

$$\text{erf}(w) = \frac{2}{\sqrt{\pi}} \int_o^w e^{-\beta^2} d\beta \quad \text{where } w = \frac{x}{2\sqrt{(\alpha\tau)}}$$

w	erf (w)
0.00	0.0000
0.01	0.0113
0.02	0.0226
0.04	0.0451
0.06	0.0676
0.08	0.0901
0.10	0.1125
0.20	0.2227
0.30	0.3286
0.40	0.4284
0.50	0.5205
0.60	0.6039
0.70	0.6778
0.80	0.7421
0.90	0.7969
1.00	0.8427
1.10	0.8802
1.20	0.9103
1.30	0.9340
1.40	0.9523
1.50	0.9661
1.60	0.9763
1.70	0.9838
1.80	0.9891
2.00	0.9953
2.20	0.9981
2.50	0.9996
3.00	1.0000

BESSEL'S FUNCTIONS (Cylindrical Functions)

x	$J_0(x)$	$J_1(x)$	$Y_0(x)$	$Y_1(x)$	$I_0(x)$	$I_1(x)$	$K_0(x)$	$K_1(x)$
0.0	1.0000	0.0000	$-\infty$	$-\infty$	1.0000	0.0000	∞	∞
0.1	0.9975	0.0499	-1.5342	-6.4590	1.003	0.0501	2.4271	9.8538
0.2	0.9900	0.0995	-1.0811	-3.3238	1.010	0.1005	1.7527	4.7760
0.3	0.9776	0.1483	-0.8073	-2.2931	1.023	0.1517	1.3725	3.0560
0.4	0.9604	0.1960	-0.6060	-1.7809	1.040	0.2040	1.1145	2.1844
0.5	0.9385	0.2423	-0.4445	-1.4715	1.063	0.2579	0.9244	1.6564
0.6	0.9120	0.2867	-0.3085	-1.2604	1.092	0.3137	0.7775	1.3028
0.7	0.8812	0.3290	-0.1907	-1.1032	1.126	0.3719	0.6605	1.0503
0.8	0.8463	0.3688	-0.0868	-0.9781	1.167	0.4329	0.5653	0.8618
0.9	0.8075	0.4059	0.0056	-0.8731	1.213	0.4971	0.4867	0.7165
1.0	0.7652	0.4401	0.0883	-0.7812	1.266	0.5652	0.4210	0.6019
1.1	0.7196	0.4709	0.1622	-0.6981	1.326	0.6375	0.3656	0.5098
1.2	0.6711	0.4983	0.2281	-0.6211	1.394	0.7147	0.3185	0.4346
1.3	0.6201	0.5220	0.2865	-0.5485	1.469	0.7973	0.2782	0.3725
1.4	0.5669	0.5419	0.3379	-0.4791	1.533	0.8861	0.2437	0.3208
1.5	0.5118	0.5579	0.3824	-0.4123	1.647	0.9817	0.2138	0.2774
1.6	0.4554	0.5699	0.4204	-0.3476	1.750	1.085	0.1880	0.2406
1.7	0.3980	0.5778	0.4520	-0.2847	1.864	1.196	0.1655	0.2094
1.8	0.3400	0.5815	0.4774	-0.2237	1.990	1.317	0.1459	0.1826
1.9	0.2818	0.5812	0.4968	-0.1644	2.128	1.448	0.1288	0.1597
2.0	0.2239	0.5767	0.5104	-0.1070	2.280	1.591	0.1139	0.1399
2.1	0.1666.	0.5683	0.5183	-0.0517	2.446	1.745	0.1008	0.1227
2.2	0.1104	0.5560	0.5208	+0.0015	2.629	1.914	0.0893	0.1079
2.3	0.0555	0.5399	0.5181	0.0523	2.830	2.098	0.07914	0.09498
2.4	0.0025	0.5202	0.5104	0.1005	3.049	2.298	0.07022	0.08372
2.5	-0.0484	0.4971	0.4981	0.1459	3.290	2.517	0.06352	0.07389
2.6	-0.0968	0.4708	0.4813	0.1884	3.553	2.755	0.05540	0.06528
2.7	-0.1424	0.4416	0.4605	0.2276	3.842	3.016	0.04926	0.05774
2.8	-0.1850	0.4097	0.4359	0.2635	4.157	3.301	0.04382	0.05111
2.9	-0.2243	0.3754	0.4079	0.2959	4.503	3.613	0.03901	0.04529
3.0	-0.2601	0.3391	0.3769	0.3247	4.881	3.953	0.03474	0.04016
3.1	-0.2921	0.3009	0.3431	0.3496	5.294	4.326	0.03095	0.03563
3.2	-0.3202	0.2613	0.3070	0.3707	5.747	4.734	0.02759	0.03164
3.3	-0.3443	0.2207	0.2691	0.3879	6.243	5.181	0.02461	0.02812
3.4	-0.3643	0.1792	0.2296	0.4010	6.785	5.670	0.02196	0.02500
3.5	-0.3801	0.1374	0.1890	0.4102	7.378	6.206	0.01960	0.02224
3.6	-0.3918	0.0955	0.1477	0.4154	8.028	6.793	0.01750	0.01979
3.7	-0.3992	0.0538	0.1061	0.4167	8.739	7.436	0.01563	0.01763
3.8	-0.4026	0.0128	0.0645	0.4141	9.517	8.140	0.01397	0.01571
3.9	-0.4018	-0.0272	0.0234	0.4078	10.37	8.913	0.01248	0.01400
4.0	-0.3971	-0.0660	-0.0169	0.3979	11.30	9.759	0.01116	0.01248
4.1	-0.3887	-0.1033	-0.0561	0.3846	12.32	10.69	0.00998	0.01114
4.2	-0.3766	-0.1386	-0.0938	0.3680	13.44	11.71	0.00893	0.00994
4.3	-0.3610	-0.1719	-0.1296	0.3484	14.67	12.82	0.00799	0.00887
4.4	-0.3423	-0.2028	-0.1633	0.3260	16.01	14.05	0.00715	0.00792
4.5	-0.3205	-0.2311	-0.1947	0.3010	17.48	15.39	0.00640	0.00708
4.6	-0.2961	-0.2566	-0.2235	0.2737	19.09	16.86	0.00573	0.00633
4.7	-0.2693	-0.2791	-0.2494	0.2445	20.86	18.48	0.00513	0.00565
4.8	-0.2404	-0.2985	-0.2723	0.2136	22.79	20.25	0.00460	0.00506
4.9	-0.2097	-0.3147	-0.2921	0.1812	24.91	22.20	0.00412	0.00452
5.0	-0.1776	-0.3276	-0.3085	0.1479	27.24	24.34	0.00369	0.00404

BESSEL'S FUNCTIONS (Cylindrical Functions) (Cont.)

x	$J_0(x)$	$J_1(x)$	$Y_0(x)$	$Y_1(x)$	$I_0(x)$	$I_1(x)$	$K_0(x)$ 0.00	$K_1(x)$ 0.00
5.1	-0.1443	-0.3371	-0.3216	0.1137	29.79	26.68	3308	3619
5.2	-0.1103	-0.3432	-0.3313	0.0792	32.58	29.25	2966	3239
5.3	-0.0758	-0.3460	-0.3374	0.0445	35.65	32.08	2659	2900
5.4	-0.0412	-0.3453	-0.3402	0.0101	39.01	35.18	2385	2597
5.5	-0.0068	-0.3414	-0.3395	-0.0238	42.69	38.59	2139	2326
5.6	0.0270	-0.3343	-0.3354	-0.0568	46.74	42.33	1918	2083
5.7	0.0599	-0.3241	-0.3282	-0.0887	51.17	46.44	1721	1866
5.8	0.0917	-0.3110	-0.3177	-0.1192	56.04	50.95	1544	1673
5.9	0.1220	-0.2951	-0.3044	-0.1481	61.38	55.90	1386	1499
6.0	0.1506	-0.2767	-0.2882	-0.1750	67.23	61.34	1244	1344
6.1	0.1773	-0.2559	-0.2694	-0.1998	73.66	67.32	1117	1205
6.2	0.2017	-0.2329	-0.2483	-0.2223	80.72	73.89	1003	1081
6.3	0.2238	-0.2081	-0.2251	-0.2422	88.46	81.10	09001	09691
6.4	0.2433	-0.1816	-0.1999	-0.2596	96.96	89.03	08083	08693
6.5	0.2601	-0.1538	-0.1732	-0.2741	106.3	97.74	07259	07799
6.6	0.2740	-0.1250	-0.1452	-0.2857	116.5	107.3	06520	06998
6.7	0.2851	-0.0953	-0.1162	-0.2945	127.8	117.8	05857	06280
6.8	0.2931	-0.0652	-0.0864	-0.3002	140.1	129.4	05262	05636
6.9	0.2981	-0.0349	-0.0563	-0.3029	153.7	142.1	04728	05059
7.0	0.3001	-0.0047	-0.0259	-0.3027	168.6	156.0	04248	04542
7.1	0.2991	0.0252	0.0042	-0.2995	185.0	171.4	03817	04078
7.2	0.2951	0.0543	0.0339	-0.2934	202.9	188.3	03431	03662
7.3	0.2882	0.0826	0.0628	-0.2846	222.7	206.8	03084	03288
7.4	0.2786	0.1096	0.0907	-0.2731	244.3	227.2	02772	02953
7.5	0.2663	0.1352	0.1173	-0.2591	268.2	249.6	02492	02653
7.6	0.2516	0.1592	0.1424	-0.2428	294.3	274.2	02240	02383
7.7	0.2346	0.1813	0.1658	-0.2243	323.1	301.1	02014	02141
7.8	0.2154	0.2014	0.1872	-0.2039	354.7	331.1	01811	01924
7.9	0.1944	0.2192	0.2065	-0.1817	389.4	363.9	01629	01729
8.0	0.1717	0.2346	0.2235	-0.1581	427.6	399.9	01465	01554
8.1	0.1475	0.2476	0.2381	-0.1331	469.3	439.5	01317	01396
8.2	0.1222	0.2580	0.2501	-0.1072	515.6	483.0	01185	01255
8.3	0.0960	0.2657	0.2595	-0.0806	566.3	531.0	01066	01128
8.4	0.0692	0.2708	0.2662	-0.0535	621.9	583.7	009588	01014
8.5	0.0419	0.2731	0.2702	-0.0262	683.2	641.6	008626	009120
8.6	0.0146	0.2728	0.2715	0.0011	750.5	705.4	007761	008200
8.7	-0.0125	0.2697	0.2700	0.0280	824.4	775.5	006983	007374
8.8	-0.0392	0.2641	0.2659	0.0544	905.8	852.7	006283	006631
8.9	-0.0653	0.2559	0.2592	0.0799	995.2	937.5	005654	005964
9.0	-0.0903	0.2453	0.2499	0.1043	1094	1031	005088	005364
9.1	-0.1142	0.2324	0.2383	0.1275	1202	1134	004579	004825
9.2	-0.1367	0.2174	0.2245	0.1491	1321	1247	004121	004340
9.3	-0.1577	0.2004	0.2086	0.1691	1451	1371	003710	003904
9.4	-0.1768	0.1816	0.1907	0.1871	1595	1508	003339	003512
9.5	-0.1939	0.1613	0.1712	0.2032	1753	1658	003006	003160
9.6	-0.2090	0.1395	0.1502	0.2171	1927	1824	002706	002843
9.7	-0.2218	0.1166	0.1279	0.2287	2119	2006	002436	002559
9.8	-0.2323	0.0928	0.1045	0.2379	2329	2207	002193	002303
9.9	-0.2403	0.0684	0.0804	0.2447	2561	2428	001975	002072
10.0	-0.2459	0.0435	0.0557	0.2490	2816	2671	001778	001865

RADIATION

Description	Equation	Notation
Wave length	$\lambda = \dfrac{c}{\nu}$	λ, wave length, m
		c, speed of light $= 3\times10^{8}$ m/s
Radiant Energy	$\rho + \alpha + \tau = 1$	ν, frequency, s^{-1}
		ρ, reflectivity
Quantum theory	$E = h\nu$	α, absorbtivity
Kirchoff's Law	$\varepsilon_\lambda = \alpha_\lambda$	τ, transmissivity
	$\varepsilon_\lambda = \alpha_\lambda = \varepsilon = \alpha$	E, energy radiated
	for gray bodies	h, Planck constant
		$= 67.5\times10^{-36}$ kgf m sec
Wien's Law	$\lambda_{max} T = 2.9 \times 10^{-3}$ m K	$= 662\times10^{-36}$ Js
		ε, emissivity $= \dfrac{E}{E_b}$
Stefan-Boltzmann Law for BLACK bodies	$E_b = \sigma T^4$	ε_λ, monochromatic emissivity
Planck's Law	$E_{b\lambda} = \dfrac{C_1 \lambda^{-5}}{e^{C_2/\lambda T} - 1}$	T, temp. K
		σ, Stefan - Boltzmann constant
		$= 4.9\times10^{-8}$ kcal/m^2 h K^4
		$= 5.7\times10^{-8}$ W/m^2 K^4
Exchange between gray bodies	$Q = \dfrac{\varepsilon}{\rho} A (E_b - J)$	*$C_1 = 0.321\times10^{-15}$ kcal m^2/h
	$J = \rho G + \varepsilon E_b$	$= 0.374\times10^{-15}$ Wm^2
		$C_2 = 14.4\times10^{-3}$ mK
		E_b, emissive power of black body
		$E_{b\lambda}$, emissive power for black body per unit wave length
		Q, net heat interchange
		J, radiosity rate at which radiation leaves a given surface
		G, irradiation (incident radiation on a surface)

*The constant C_1 is to be divided by 2π for polarised radiation normal to the surface.

General Equations	$Q_{ij} = [E_{bi} - E_{bj}] F_{ij} A_i$ $$F_{ij} A_i = F_{ji} A_j$$	Q_{ij}, Net beat exchange from i to j E_{bi}, emissive power of body i F_{ij}, geometric shape factor of surface i with respect to j
Geometry	Shape factor F_{ij} for black bodies	for gray bodies $A_i F_{ij}$
E qual and parallel squares, rectangles and discs	Refer pages 77 to 79 and 82	
Adjacent rectangles at right angles	Refer pages 80 & 81	
Opening in walls	Refer page 82	
Surface element and a parallel rectangular surface	Refer page 89	
Surface element and a per-pendicular rectangular surface	Refer page 84	$1 - \left[\dfrac{1 - \varepsilon_j}{A_j \varepsilon_j} + \dfrac{1}{A_i F_{ij}} + \dfrac{1 - \varepsilon_i}{A_i \varepsilon_i} \right]$
Differential spherical surface and a rectangular surface	Refer page 84	
A plane with parallel rows of tubes	Refer pages 85 & 86	
Parallel infinite planes	1	
Totally enclosed small body i [small compared with a large enclosing body j]	1	
Totally enclosed large body	1	
Concentric infinite long cylinders, concentric spheres	1	

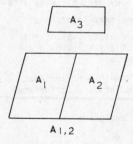

$A_{1,2}$

1) $F_{3-1,2} = F_{3-1} + F_{3-2}$

2) $A_3 F_{3-1,2} = A_3 F_{3-1} + A_3' F_{3-2}$

3) $A_{1,2} F_{1,2-3} = A_1 F_{1-3} + A_2 F_{2-3}$

Ratio of black body radiant energy upto wavelength λ to the total emissive power of a black body at the same temperature.

λ .T μ K	$\dfrac{E_{b,o-\lambda,T}}{\sigma T^4}$	λ .T μ K	$\dfrac{E_{b,o-\lambda,T}}{\sigma T^4}$
800	0.000	6111	0.7474
900	0.0001	6222	0.7559
1000	0.0003	6333	0.7643
1111	0.0009	6444	0.7724
1222	0.0025	6555	0.7802
1333	0.0053	6666	0.7876
1444	0.0098	6777	0.7947
1555	0.0164	6888	0.8015
1666	0.0254	7000	0.8081
1777	0.0368	7111	0.8144
1888	0.0506	7222	0.8204
2000	0.0667	7333	0.8262
2111	0.0850	7444	0.8317
2222	0.1051	7555	0.8370
2333	0.1267	7666	0.8421
2444	0.1496	7777	0.8470
2555	0.1734	7888	0.8517
2666	0.1979	8000	0.8563
2777	0.2229	8111	0.8606
2888	0.2481	8222	0.8648
3000	0.2733	8333	0.8688
3111	0.2983	8889	0.8868
3222	0.3230	9444	0.9017
3333	0.3474	10000	0.9142
3444	0.3712	10555	0.9247
3555	0.3945	11111	0.9335
3666	0.4171	11666	0.9411
3777	0.4391	12222	0.9475
3888	0.4604	12777	0.9531
4000	0.4809	13333	0.9589
4111	0.5007	13888	0.9621
4222	0.5199	14444	0.9657
4333	0.5381	15000	0.9689
4444	0.5558	15555	0.9718
4555	0.5727	16111	0.9742
4666	0.5890	16666	0.9765
4777	0.6045	22222	0.9881
4888	0.6195	27777	0.9941
5000	0.6337	33333	0.9963
5111	0.6474	38888	0.9931
5222	0.6606	44444	0.9987
5333	0.6731	50000	0.9990
5444	0.6851	55555	0.9992
5555	0.6966		
5666	0.7076		
5777	0.7181		
5888	0.7282		
6000	0.7378		

GAS RADIATION

Condition of gas or mixture of gases	*Emissivity, ε_g

1. Carbon dioxide
    i) at a total pressure of one atm.
 	
    ii) at total pressures other than one atm.
	Refer chart on page 87 for ε_{CO_2}
	Refer bottom chart on page 87 for correction factor, C_p, by which ε_{CO_2} has to be multiplied

2. Water Vapour
    i) at a total pressure of one atm. and near zero partial pressure
    ii) at a total pressure of one atm. and other partial pressures
 	
    iii) at total pressures other than one atm.
	Refer page 88 for ε_{H_2O}
	
	Refer bottom chart on page 89 for correction factor β by which ε_{H_2O} has to be multiplied
	Refer bottom chart on page 88 for correction factor C_p
	$$\varepsilon_g = \varepsilon_{H_2O}\ \beta\ C_p$$

3. Mixture of Water Vapour and carbon dioxide	Refer to top chart on page 89 $$\varepsilon_g = \varepsilon_{H_2O} + \varepsilon_{CO_2} - \Delta\varepsilon_g$$ ε_{H_2O} and ε_{CO_2} have to be determined for the particular cases.

*For non-luminous gas radiation layers of different shapes, an equivalent thickness given on page 85 should be used for ℓ in the charts on pp. 87 to 89.

EMISSIVITIES ε_n OF THE RADIATION IN THE DIRECTION OF THE NORMAL TO THE SURFACE AND ε OF THE TOTAL HEMISPHERICAL RADIATION FOR VARIOUS MATERIALS

In the instances where the exact measurements are not given, take for bright metal surfaces an average ratio $\varepsilon/\varepsilon_n = 1.2$, and for other substances with smooth surfaces $\varepsilon/\varepsilon_n = 0.95$, and in the case of rough surfaces use $\varepsilon/\varepsilon_n = 0.98$.

Surface	Temperature °C	ε_n	ε
Gold, polished	130	0.018	-
Gold, polished	400	0.022	-
Silver, pure polished	200-600	0.020-0.030	-
Copper, polished	20	0.030	-
Copper, lightly oxidized	20	0.037	-
Copper, scraped	20	0.070	-
Copper, black oxidized	20	0.780	-
Copper, oxidized	131.09	0.760	0.725
Aluminium, bright rolled	170	0.039	0.049
Aluminium, bright rolled	500	0.050	-
Aluminium, paint	100	0.20-0.40	-
Silumin, cast polished	150	0.186	-
Nickel, bright matte	100	0.041	0.046
Nickel, polished	100	0.045	0.053
Chrome, polished	150	0.058	0.071
Iron, bright etched	150	0.128	0.158
Iron, bright abrased	20	0.240	-
Iron, red rusted	20	0.610	-
Iron, hot rolled	20	0.770	-
Iron, hot rolled	130	0.600	-
Iron, hot cast	100	0.800	-
Iron, heavily rusted	0	0.850	-
Iron, heat-resistant oxidized	80	0.613	-
Iron, heat-resistant oxidized	200	0.639	-
Zinc, gray oxidized	20	0.23-0.28	-
Lead, gray oxidized	20	0.28	-
Bismuth, bright	80	0.340	0.366
Corundum, emery rough	80	0.855	0.84
Clay, fired	70	0.91	0.86
Lacquer, white	100	0.925	-
Red lead	100	0.930	-
Enamel, lacquer	20	0.85-0.95	-
Lacquer, black matte	80	0.970	-
Bakelite lacquer	80	0.935	-
Brick, mortar, plaster	20	0.930	-
Porcelain	20	0.92-0.94	-
Glass	90	0.940	0.876
Ice, smooth and water	0	0.966	0.918
Ice, rough crystals	0	0.985	-
Water glass	20	0.960	-
Paper	95	0.920	0.890
Wood, beech	70	0.935	0.910
Tar paper	20	0.930	-
Manganin, bright rolled	118.33	0.048	0.057

1 White fire clay   5 Porcelain
2 Asbestos   6 Cencrete
3 Cork   7 Aluminium
4 Wood   8 Graphite

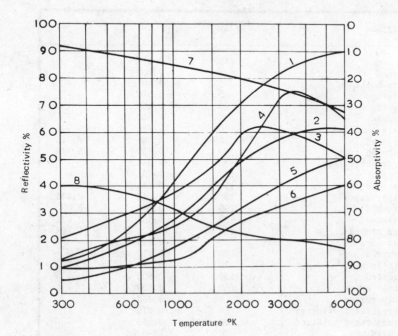

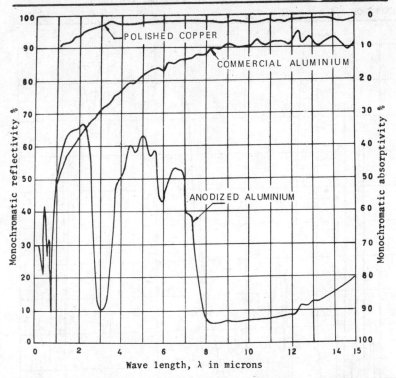

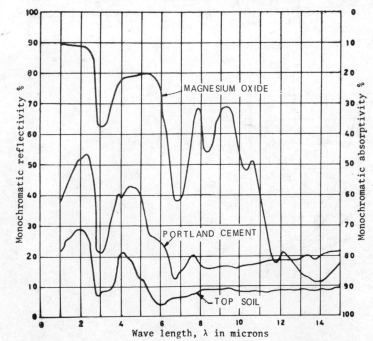

75

ABSORPTION BANDS OF WATER VAPOUR AT ATMOSPHERIC PRESSURE AND 140°C FOR A THICKNESS OF 1.1 m

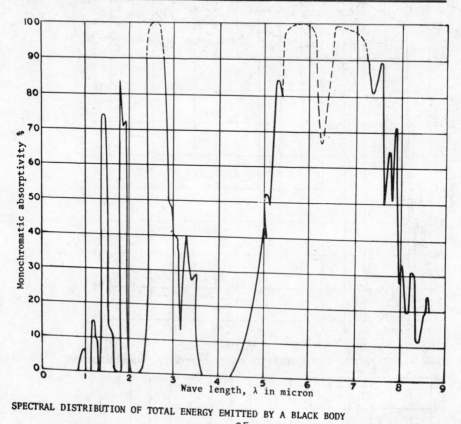

SPECTRAL DISTRIBUTION OF TOTAL ENERGY EMITTED BY A BLACK BODY

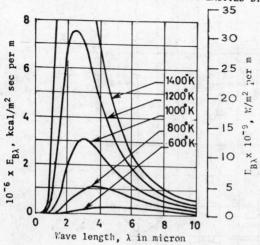

SHAPE FACTORS FOR EQUAL AND PARALLEL SQUARES, RECTANGLES AND DISCS
(THE CURVES LABELLED 5, 6, 7 and 8 ALLOW FOR CONTINUOUS VARIATION
IN THE SIDE WALL TEMPERATURES FROM TOP TO BOTTOM)

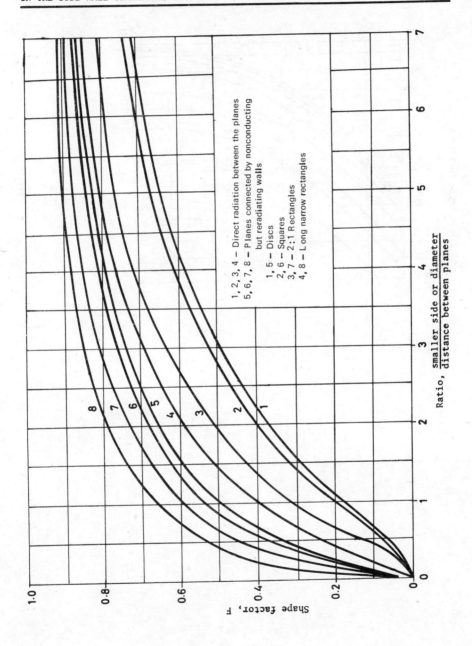

1, 2, 3, 4 – Direct radiation between the planes
5, 6, 7, 8 – Planes connected by nonconducting
 but reradiating walls

1, 5 – Discs
2, 6 – Squares
3, 7 – 2:1 Rectangles
4, 8 – Long narrow rectangles

Shape factor, F

Ratio, $\dfrac{\text{smaller side or diameter}}{\text{distance between planes}}$

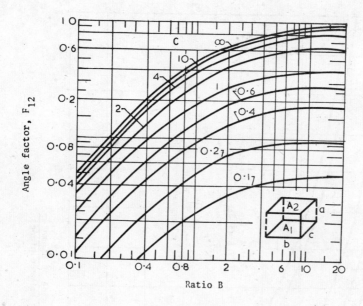

CONFIGURATION FACTORS : PARALLEL SURFACES

$X = \dfrac{b}{c}$; $Y = \dfrac{a}{c}$

X→ / Y↓	.1	.2	.4	.6	1.0	2.0	4.0	10.0	∞
0.1	.00316	.00626	.01207	.01715	.02492	.03514	.04209	.04671	.04988
0.2	.00626	.01240	.02392	.03398	.04941	.06971	.08353	.09270	.09901
0.4	.01207	.02391	.04614	.06560	.09554	.13513	.16219	.18021	.19258
0.6	.01715	.03398	.06560	.09336	.13627	.19342	.23171	.25896	.27698
1.0	.02492	.04941	.09554	.13627	.19982	.28588	.34596	.38638	.41421
2.0	.03514	.06971	.13513	.19341	.28588	.41525	.50899	.57361	.61803
4.0	.04210	.08353	.16219	.23271	.34596	.50899	.63204	.71933	.78078
6.0	.04463	.08859	.17209	.24712	.36813	.54421	.67954	.77741	.84713
10.0	.04671	.09272	.18021	.25896	.38638	.57338	.71933	.82699	.90499
20.0	.04829	.09586	.18658	.26795	.40026	.59563	.74990	.86563	.95125
∞	.04988	.09902	.19258	.27698	.41421	.61803	.78078	.90499	1.00000

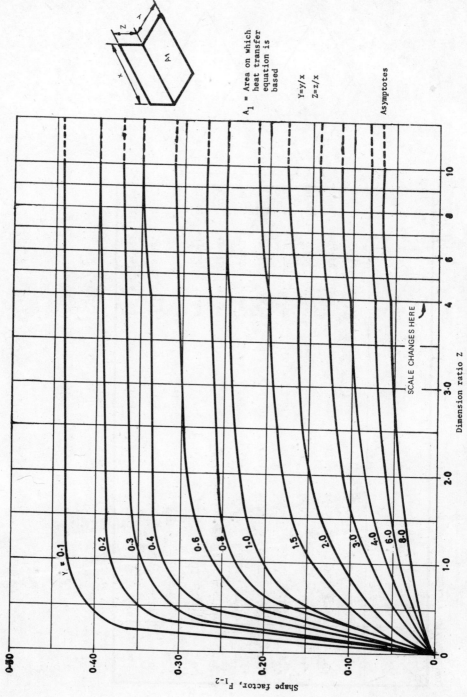

A_1 = Area on which heat transfer equation is based

Y=y/x

Z=z/x

Asymptotes

Shape factor, F_{1-2}

Dimension ratio Z

Shape factor for adjacent rectangles in perpendicular planes

Y = 0.1

0.2

0.3

0.4

0.6

0.8

1.0

1.5

2.0

3.0

4.0

6.0

8.0

SCALE CHANGES HERE

CONFIGURATION FACTORS : PERPENDICULAR SURFACES

$Z = z/x$
$Y = y/x$

Z ↓ \\ Y →	.02	.05	.1	.2	.4	.6	1.0	2.0	4.0	6.0	10.0	20.0
.05	0.39908	.28738	.18601	.10584	.05606	.03799	.02304	.01158	.00580	.00388	.00232	.00116
.1	.44375	.37202	.28189	.18108	.10215	.07048	.04326	.02188	.01097	.00732	.00439	.00220
.2	.46615	.42337	.36216	.27104	.17147	.12295	.07744	.03971	.02000	.01335	.00802	.00401
.4	.47725	.44852	.40859	.34295	.25032	.19206	.12770	.06757	.03434	.02296	.01380	.00691
.6	.47943	.45587	.42290	.36884	.28809	.23147	.16138	.08829	.04556	.03040	.01829	.00916
1.0	.48138	.46073	.43251	.38719	.31924	.26896	.20004	.11643	.06131	.04130	.02492	.01249
2.0	.48239	.46327	.43756	.39711	.33784	.29429	.23285	.14930	.08365	.05731	.03491	.01757
4.0	.48267	.46397	.43898	.39994	.34339	.30238	.24522	.16731	.10136	.07184	.04484	.02285
6.0	.48273	.46411	.43925	.40048	.34447	.30399	.24783	.17193	.10776	.07822	.04998	.02587
10.0	.48275	.46418	.43939	.40076	.34503	.30482	.24921	.17455	.11210	.08331	.05502	.02938
20.0	.48276	.46421	.43943	.40089	.34528	.30518	.24980	.17571	.11427	.08624	.05876	.03302
∞	.48277	.46422	.43947	.40092	.34535	.30530	.25000	.17611	.11505	.08738	.06053	.03578

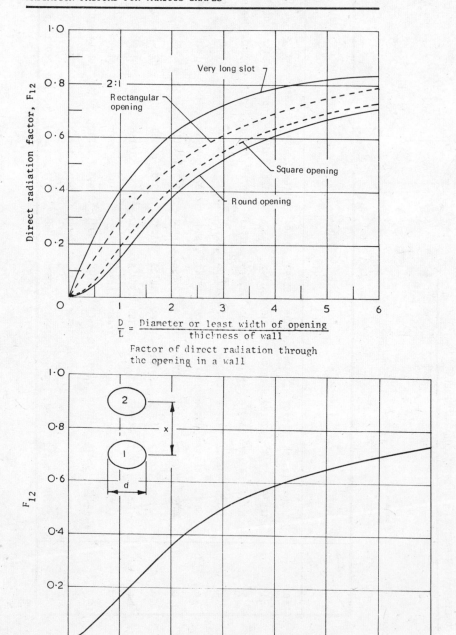

$$\frac{D}{L} = \frac{\text{Diameter or least width of opening}}{\text{thickness of wall}}$$

Factor of direct radiation through
the opening in a wall

Radiation shape factor F_{12} for radiation
between parallel discs

82

SHAPE FACTOR FOR A SURFACE ELEMENT AND A RECTANGULAR SURFACE PARALLEL TO IT

L_1 and L_2 are sides
of rectangle
D is distance from dA
to rectangle
F is shape factor

$F = 0.02$
0.03
0.04
0.05
0.06
0.08
0.10
0.12
0.14
0.16
0.18
0.20
0.22
0.24

D/L_1 Dimension ratio

D/L_2 Dimension ratio

83

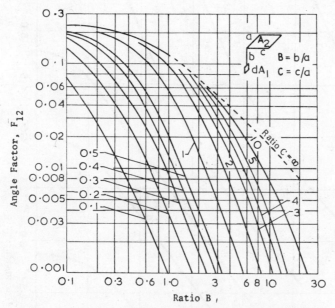

Angle Factor for a system of a differential
surface and a finite rectangular surface
perpendicular to it

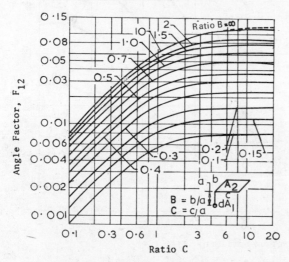

Angle Factor for a system of a differential
spherical surface and a finite rectangular
surface

84

SHAPE FACTOR FOR A PLANE AND ONE OR TWO ROWS OF TUBES ABOVE AND PARALLEL TO IT

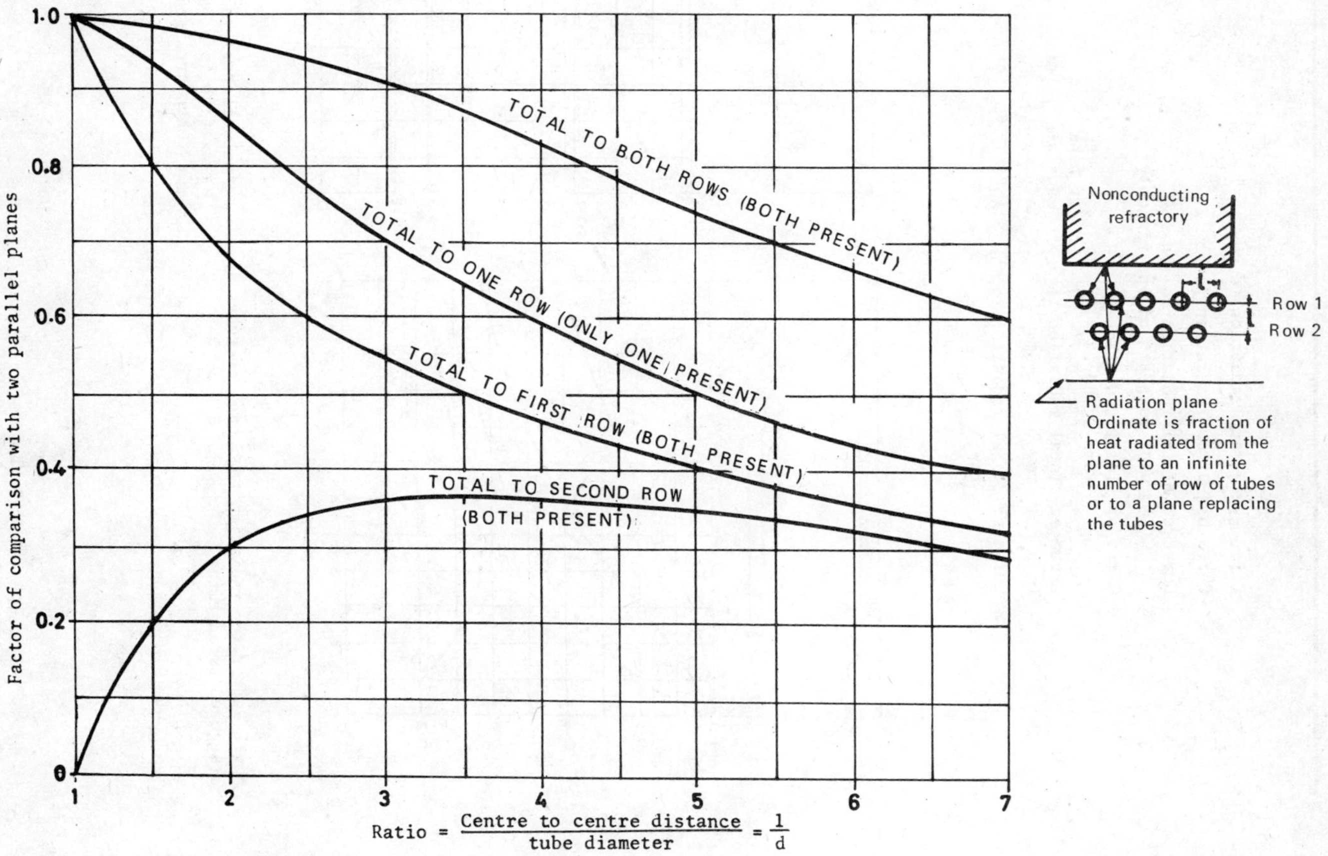

Nonconducting refractory

Row 1
Row 2

Radiation plane
Ordinate is fraction of heat radiated from the plane to an infinite number of row of tubes or to a plane replacing the tubes

TOTAL TO BOTH ROWS (BOTH PRESENT)

TOTAL TO ONE ROW (ONLY ONE PRESENT)

TOTAL TO FIRST ROW (BOTH PRESENT)

TOTAL TO SECOND ROW (BOTH PRESENT)

Factor of comparison with two parallel planes

$$\text{Ratio} = \frac{\text{Centre to centre distance}}{\text{tube diameter}} = \frac{l}{d}$$

Curve a: Direct radiation to row 1

Curve b: Direct radiation to row 2

Curve c: Total radiation to row 1 if bank consists of one row only

Curve d: Total radiation to rows 1 and 2 ⎫

Curve e: Total radiation to row 1 ⎬ if bank consists of two rows

Curve f: Total radiation to row 2 ⎭

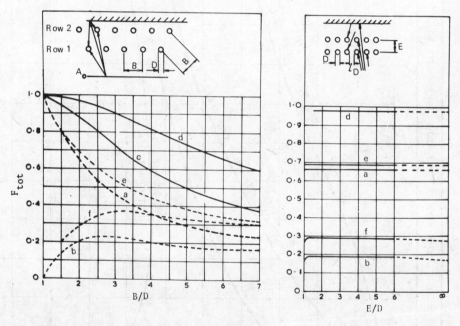

Angle Factor for radiation to a bank of tubes in staggered arrangement

Angle factors for radiation to a bank of tubes arranged in line

86

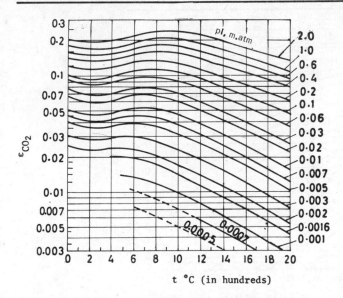

$\varepsilon_{CO_2} = f(t, pl) = $ emissivity of carbon dioxide

$P_{total} = 1$ atm and near zero partial pressure

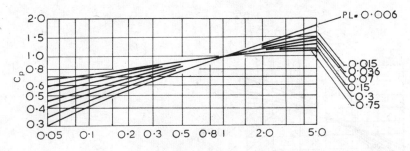

P_T, total pressure atm

Factor C_p for correcting emissivity of CO_2 at
1 atm total pressure to emissivity at P_T atm

For ℓ refer table on page 90

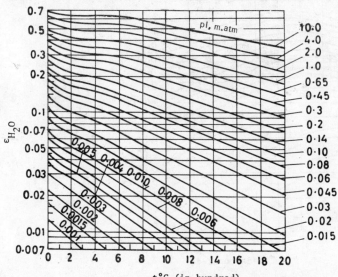

ϵ_{H_2O} = f(t, pl) = emissivity of water vapour

P_{total} = 1 atm and near zero partial pressure

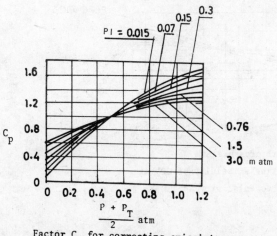

Factor C_p for correcting emissivity
of water vapour to values P and P_T
other than 0 and 1 atm

For ℓ refer table on page 90

88

CORRECTION TO EMISSIVITY OF MIXTURE OF WATER VAPOUR AND CARBON DIOXIDE AND
WATER VAPOUR ACCOUNTING FOR PARTIAL PRESSURE

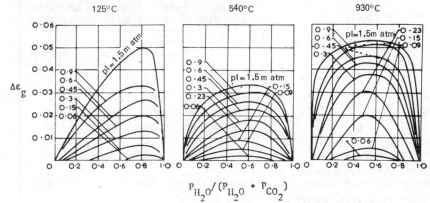

$$P_{H_2O}/(P_{H_2O} + P_{CO_2})$$

Correction to the emissivity of a mixture of water
vapour and carbon dioxide

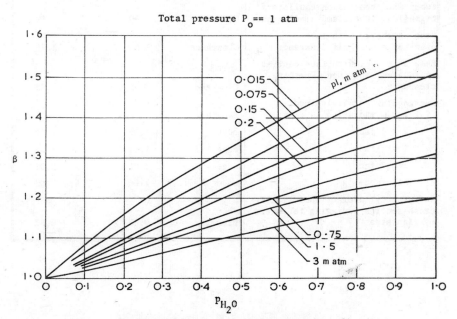

Correction factor β for the emissivity of water
vapour accounting for partial pressure

89

EQUIVALENT THICKNESS, ℓ FOR NON-LUMINOUS GAS RADIATION LAYERS OF DIFFERENT SHAPES

Shape	Characteristic dimension Z	Factor by which Z is to be multiplied to give equivalent ℓ for hemispherical radiation	
		Calculated by various workers	3.4 x (volume/area)
Sphere	Diameter	0.60	0.57
Cube	Side	0.60	0.57
Infinite cylinder radiating to walls	Diameter	0.90	0.85
Infinite cylinder radiating to centre of base	Diameter	0.90	0.85
Cylinder, height = diameter, radiating to whole surface	Diameter	0.60	0.57
Cylinder (ht = diam) radiating to centre of base	Diameter	0.77	0.57
Space between infinite parallel planes	Distance apart	1.80	1.70
Space outside infinite bank of tubes with centres on equilateral triangles, tube diameter=clearance	Clearance	2.80	2.89
Same as above, with tube diameter = one half clearance	Clearance	3.80	3.78
Same as above, with tube centres on squares, and tube diameter= clearance	Clearance	3.50	3.49
Rectangular parallelepiped, 1 x 2 x 6 radiating to :	Shortest		
2 x 6 face	edge	1.06	1.01
1 x 6 face	edge	1.06	1.05
1 x 2 face	edge	1.06	1.01
all faces	edge	1.06	1.02
Infinite cylinder of semicircular cross-section radiating to centre of flat side	Diameter	0.63	0.52

BOUNDARY LAYER

δ, Thickness of boundary layer (δ_h or δ_T)

δ_h, Hydrodynamic boundary layer thickness

δ_T, Thermal boundary layer thickness

C_f, Coefficient of friction = $\dfrac{2\tau_w g_o}{\rho u^2_\infty}$; τ_w, wall shear stress

h, heat transfer coefficient

$$\frac{h}{\rho u_\infty C_p} = St; \qquad \frac{hx}{k} = Nu_x; \qquad \frac{u_\infty x}{\nu} = Re_x; \qquad \frac{C_p \mu}{k} = Pr$$

$$\text{Also } St = \frac{Nu}{RePr}$$

1. FLAT PLATE — Laminar region $[\delta_T \le \delta_h]$

$\dfrac{u}{u_\infty}$ or $\dfrac{T}{T_\infty}$	$\dfrac{\delta}{x} Re_x^{0.5}$	$C_{fx} Re_x^{0.5}$	$\dfrac{Nu_x}{Re_x^{0.5} Pr^{0.33}}$
Exact solution from diff. eqn.	5	0.664	0.332
$\dfrac{y}{\delta}$	3.46	0.578	0.289
$\dfrac{3}{2}(\dfrac{y}{\delta}) - \dfrac{1}{2}(\dfrac{y}{\delta})^3$	4.64	0.646	0.323
$Sin\left(\dfrac{\pi}{2} \dfrac{y}{\delta}\right)$	4.80	0.654	—

DIMENSIONLESS GROUPS:

Groups	Symbol	Name	Significance
hx/k_s	Bi	Biot Number	Surface convection resistance / Internal conduction resistance
$\Delta P/\rho u^2$	Eu	Euler Number	Pressure force/Inertia force
$\dfrac{\alpha t}{s^2}$	Fo	Fourier Number	Characteristic body dimension / Temperature wave penetration depth in time, τ
u^2/gL	Fr	Froude Number	Inertia force/gravity force
$\dfrac{k}{ux\rho c}$	$Gz = (\frac{L}{x})/RePr$	Graetz Number	Heat transfer by conduction / Heat transfer by convection (with entrance region consideration)
$(g\beta\Delta TL^3 \rho^2)/\mu^2$	Gr	Grashof Number	(Buoyance force)/(Viscous force) (Inertia force)/(Viscous force)
λ/L	Kn	Knudson Number	Mean free path / Characteristic body dimension
α/D	Le	Lewis Number	Heat diffusivity / Mass diffusivity
$\dfrac{hx}{k}$	Nu	Nusselt Number	Ratio of temperature gradients by conduction and convection at the surface
$ux\rho c/k$	Pe = RePr	Peclet Number	Heat transfer by convection / Heat transfer by conduction
$\dfrac{C_p \mu}{k}$	Pr	Prandtl Number	Molecular diffusivity of momentum / Molecular diffusivity of heat
$g\beta\Delta TL^3 \dfrac{\rho^2 C_p}{\mu k}$	Ra = GrPr	Rayleigh Number	Refer Pr and Gr
$\dfrac{ux}{\nu}$	Re	Reynolds Number	Inertia force/viscous force
$\mu/\rho D$	Se	Schmidt Number	Molecular diffusivity of momentum / Molecular diffusivity of mass
Kx/D	Sh	Sherwood Number	Ratio of concentration gradients at the boundary (by diffusion and by convection)
$\dfrac{h}{C_p \rho u}$	$St = \dfrac{Nu}{RePr}$	Stanton Number	Wall heat transfer rate / Heat transfer by convection

CONVECTION (Cont.)

2. FLAT PLATE:

	$\dfrac{\delta_h}{x}$	displacement thickness δ^*	momentum thickness δ_i	$\dfrac{\delta_T}{\delta_h}$	Wall shear stress τ_w	Total drag on one side D	†Coefficient of drag C_D	Nu_x	St_x
LAMINAR	$\dfrac{4.64}{Re_x^{0.5}}$	$\approx \dfrac{\delta_h}{3}$	$\approx \dfrac{\delta_h}{7}$	$\dfrac{0.9744}{Pr^{0.333}}$	$0.323\,\dfrac{\rho u_\infty^2}{g_o}\left[\dfrac{1}{Re_x^{0.5}}\right]$	$0.6444\,\dfrac{\ell b \rho u_\infty^2}{g_o}\left[\dfrac{1}{Re_\ell^{0.5}}\right]$	$1.328\,Re_\ell^{-0.5}$	$\dfrac{0.323\,Re_x^{0.5}}{Pr^{-0.333}}$	$\dfrac{0.323\,Re_x^{-0.5}}{Pr^{0.667}}$
TURBULENT	$\dfrac{0.376}{Re_x^{0.2}}$	$\approx \dfrac{\delta_h}{8}$	$\dfrac{7}{72}\left(\delta_h\right)$	—	$0.0296\,\dfrac{\rho u_\infty^2}{g_o}\left[\dfrac{1}{Re_x^{0.2}}\right]$	$0.036\,\dfrac{\ell b \rho u_\infty^2}{g_o}\left[\dfrac{1}{Re_\ell^{0.2}}\right]$	$0.072\,Re_\ell^{-0.2}$	$\dfrac{0.0296\,Re_x^{0.8}}{Pr^{-0.333}}$	$\dfrac{0.0296\,Re_x^{-0.2}}{Pr^{0.667}}$

The transition from laminar to turbulent layer occurs for Re lying between 0.5×10^6 and 1×10^6.

ℓ, length of plate; b, breadth of plate; †, C_D for other geometries, refer pages 106 & 107.

Average Nusselt number over the length ℓ of the plate

$$Nu = 0.664\,Pr^{0.33}\,Re_\ell^{0.5} \quad \text{(laminar)}$$

$$Nu = 0.037\,Pr^{0.333}\,Re_\ell^{0.8} \quad \text{(turbulent)}$$

3. UNIVERSAL VELOCITY PROFILE

General Relationship: $u^+ = u^+ (y^+)$

$$u^+ = \frac{u}{(\tau_w g_o/\rho)^{\frac{1}{2}}}$$

$$y^+ = y \left[\frac{(\tau_w g_o/\rho)^{\frac{1}{2}}}{\nu}\right]$$

Formula	Range of validity	due to
$u^+ = y^+$ $u^+ = 2.5\ln y^+ + 5.5$	$0 \le y^+ \le 11.5$ $y^+ > 11.5$	Prandtl & Taylor
$u^+ = y^+$ $u^+ = 5\ln y^+ + 3.05$ $u^+ = 2.5\ln y^+ + 5.5$	$0 \le y^+ \le 5$ $5 \le y^+ \le 30$ $y^+ > 30$	Von Karman
$y^+ = u^+ + A [e^{Bu^+} - 1 - Bu^+ - \frac{1}{2}(Bu^+)^2 - \frac{1}{6}(Bu^+)^3]$ A = 0.1108 B = 0.4	for all y^+ Refer page 104	Spalding

CONVECTION (contd.) : ORDER OF MAGNITUDE OF HEAT TRANSFER COEFFICIENT, h

	kcal/m^2 h °C	W/m^2 K
Flowing gases	10 - 240	10 - 280
Flowing liquids	150 - 4900	170 - 5700
Flowing liquid metals	4900 - 244000	5700 - 284000
Gases (natural convection)	4 - 24	5 - 28
Boiling liquids	1000 - 244000	1000 - 284000
Condensing vapours	2400 - 24400	2840 - 28400

CONVECTION : FORCED CONVECTION

FORCED CONVECTION : Note: Fluid properties are to be taken at bulk mean temperature unless otherwise specified.

$Q = hA\,\Delta T$

Flow conditions	Formula	Range of validity	Notations
SMOOTH TUBES:			$h = \dfrac{k}{d}\,Nu$
i. Fully developed turbulent flow	$Nu = 0.023\ Re^{0.8}\ Pr^{n}$	$n = 0.4$ for heating the fluid $n = 0.3$ for cooling the fluid	Nu, Nusselt Number Re, Reynolds Number Pr, Prandtl Number μ_w, abs. Viscosity of fluid at wall temperature
ii. Same as above corrected for variations in fluid properties	$Nu = 0.027\ Re^{0.8}\ Pr^{0.33}\ [\tfrac{\mu}{\mu_w}]^{0.14}$		
iii. For entrance region a. See also chart on page 103 b. For configurations other than tubes see charts on pages 102 and 103	$Nu = 0.036\ Re^{0.8}\ Pr^{0.33}\ (\tfrac{d}{L})^{0.055}$	$10 < \dfrac{L}{d} < 400$	L, length of tube In the chart: X, entrance length D_e, equivalent diameter $= 4\,\dfrac{A}{P}$
iv. Fully developed LAMINAR flow at constant wall temperature	$Nu = 3.66 + \dfrac{0.00668\ (\tfrac{d}{L})\ Re\ Pr}{1 + 0.04\ [(\tfrac{d}{L})\ Re\ Pr\,]^{0.67}}$ *also* $Nu = 1.86\ (Re\ Pr)^{0.33}\ (\tfrac{d}{L})^{0.33}\ (\tfrac{\mu}{\mu_w})^{0.14}$	$Re\ Pr\ (\tfrac{d}{L}) > 10$	A, flow area P, wetted perimeter D, tube diameter

CONVECTION : FORCED CONVECTION (Cont.)

Flow conditions	Formula	Range of validity
For LAMINAR and TURBULENT for non-circular sections replace D by 4R where $R = \dfrac{\text{cross sectional area}}{\text{wetted diameter}}$	$f = \dfrac{64}{Re}$ $St\,(Pr)^{0.67} = \dfrac{f}{8}$ *also* $St = \dfrac{f}{8} = \dfrac{\tau g_0}{\rho V^2}$	$Re < 2000$ for $Pr \approx 1$ Use values of ε in chart, page 105 to determine f

Roughness table (right-hand column):

	εmm
Riveted steel	1 to 10
Concrete	0.3 to 3
Wood stave	0.2 to 1
Cast iron	0.25
Galvanized iron	0.15
Asphalted cast iron	0.12
Commercial steel or wrought iron	0.05
Drawn tubing	0.0015

FLOW ACROSS CYLINDERS AND SPHERES (fluid properties at mean film temperature T_f)

$$Nu = C\left[\frac{u_\infty d}{\nu}\right]^n$$

$$T_f = 0.5\,[T_w + T_\infty]$$

Re	n	C (for gases)	C (for liquids)
0.4 to 4	0.330	0.891	$0.989\ Pr^{0.333}$
4 to 40	0.385	0.821	$0.911\ Pr^{0.333}$
40 to 4000	0.466	0.615	$0.683\ Pr^{0.333}$
4000 to 40000	0.618	0.174	$0.193\ Pr^{0.333}$
40000 to 400000	0.805	0.0239	$0.0266\ Pr^{0.333}$

FLOW OF GASES OVER SPHERES (properties at T_f)

$$Nu = 0.37\ Re^{0.6}$$

$25 < Re < 100000$

FLOW OF LIQUIDS PAST SPHERES (properties at T_f)

$$Nu\,(Pr)^{-0.3} = 0.97 + 0.68\ Re^{0.5}$$

$1 < Re < 2000$

For TEMPERATURE PROFILES, refer charts on pages 110 and 111.

CONVECTION : FORCED CONVECTION (Cont.)

Flow conditions	Formula	Range of validity	Notations
FLOW OF OIL OR WATER PAST SPHERES (properties at T_∞)	$Nu\,(Pr)^{-0.3}\left(\dfrac{\mu_w}{\mu}\right)^{0.25} = 1.2 + 0.53\,Re^{0.54}$ $1 < Re < 200000$		For values of C and n refer to table on page 101
FLOW ACROSS BANKS OF TUBES (properties at T_f) In line & Staggered	$Nu = C\,Re^n$ $$\Delta P = \frac{f\,G_{max}^2\,N}{2.09\times10^8\rho_0}\left[\frac{\mu_w}{\mu_b}\right]^{0.14}$$ *For in line* $$f = \left\{0.044 + \frac{0.08\,\frac{S_p}{d}}{\left(\frac{S_n-1}{d}\right)^{0.43+1.13\frac{d}{S_p}}}\right\} Re_{max}^{-0.15}$$ *For staggered* $$f = \left\{0.25 + \frac{0.118}{\left(\frac{S_n-1}{d}\right)^{1.08}}\right\} Re_{max}^{-0.16}$$		f, Friction factor ΔP, Pressure drop G_{max}, Mass velocity at minimum area of flow N, Number of transverse rows μ_b, Abs. viscosity at bulk mean temperature μ_w, Abs. viscosity at wall temperature S_p, Pitch of column of tubes S_n, Pitch of row of tubes
FLOW OF LIQUID METALS i) Turbulent flow through tubes and uniform heat flux at the wall	$Nu = 0.625\,[Re\,Pr]^{0.4}$	$100 < Re\,Pr < 10000$ and $\left(\dfrac{L}{d}\right) > 60$	
ii) For tubes with constant wall temperature	$Nu = 7 + 0.025\,[Re\,Pr]^{0.8}$	$Re\,Pr > 100$ and $\left(\dfrac{L}{d}\right) > 60$	

Equation	Formula	Notation
	1. LIQUIDS :	
ERGUN eqn.	$Re_p = \dfrac{D U_{bs}\,\rho}{\mu(1-\varepsilon)}$	Re_p, Reynold's Number of packed bed
ERGUN eqn.	$f_p = \dfrac{g\,D H_f\,\varepsilon^3}{L\,U_{bs}^2(1-\varepsilon)}$	D, effective particle diameter $= \dfrac{6}{S_v}$
	also	S_v, specific surface of a particle $= S_p/v_p$
KOZENY-CARMAN eqn.	$f_p = \dfrac{150}{Re_p}$ for $Re_p < 1$	S_p, surface area of particle
		v_p, volume of a particle
BURKE-PLUMMER eqn.	$f_p = 1.75$ for $Re_p > 2500$	U_{bs}, superficial velocity, based on the area of an equivalent empty container $= \varepsilon U_b$
ERGUN eqn.	$f_p = \dfrac{150}{Re_p} + 1.75$ for $1 < Re_p < 2500$	U_b, average interstitial velocity of fluid
		ε, void fraction of porosity
	$S_{vm} = \Sigma x_i S_{vi}$	ρ, density of the fluid
		μ, abs. viscosity of the fluid
	$D_m = \dfrac{6}{S_{vm}}$	g, gravitational constant
		H_f, head lost due to friction
	$= \dfrac{1}{\Sigma(x_i/D_i)}$	L, length of packed bed
		f_p, friction factor
		S_{vm}, mean specific surface
		x_i, volume fraction of particles of same size
		D_i, effective particle diameter of size i
		D_m, mean effective diameter

PACKED BEDS (Cont.)

Equation	Formula	Notation
	2. GASES : Pressure drop, $$\Delta p = \frac{2 f_p \, L \, G^2}{g_o \, D \, \rho_{av}}$$ for $\frac{\Delta p}{p} < 0.1$	G, mass velocity of the gas per unit area ρ_{av}, density of gas at the arithmetic mean pressure g_o = 9.81 in MKS = 1 in S1 p, pressure of gas at entry
ECKERT eqn.	3, HEAT TRANSFER : $$Nu_p = \frac{hD}{k} = 0.8 \, Re_p^{0.7} \, Pr^{0.333}$$	h, heat transfer coeff. Nu_p, Nusselt Number for packed bed D, effective particle diameter k, thermal conductivity of fluid
	4. MASS TRANSFER : $$j_d = 1.82 \, Re_p^{-0.51} \text{ for } Re_p < 350$$ $$= 0.989 \, Re_p^{-0.41} \text{ for } Re_p > 350$$	j_d, COLBURN j factor $= \frac{h_D}{u_\infty} Sc^{0.667}$ h_D, convective mass transfer coefficient Sc, Schmidt number

NOTE: $Nu = C\, Re^n$
The values of C given in table below
are for air or gases. For liquids
multiply these tabulated values of C
only by $1.11\, Pr^{0.33}$

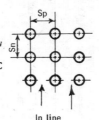

In line

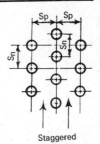

Staggered

FOR TUBE BANKS OF 10 ROWS OR MORE

Arrangement $\frac{Sp}{D}$		S_n/D							
		1.25		1.5		2.0		3.0	
		C	n	C	n	C	n	C	n
In Line	1.25	.348	.592	.275	.608	.100	.704	.0633	.752
	1.50	.367	.586	.250	.620	.101	.702	.0678	.744
	2.00	.418	.570	.299	.602	.229	.632	.1980	.648
	3.00	.290	.601	.357	.584	.374	.581	.2860	.608
Staggered	0.6	..	..	..	..	..	..	.213	.636
	0.9	..	..	..	..	.446	.571	.401	.581
	1.0	..	..	.497	.558	..	..	..	..
	1.125	..	..	..	..	.478	.565	.518	.560
	1.25	.518	.556	.505	.554	.519	.556	.522	.562
	1.5	.451	.568	.460	.562	.452	.568	.488	.568
	2.0	.404	.572	.416	.568	.482	.556	.449	.570
	3.00	.310	.592	.356	.580	.440	.562	.421	.574

Ratio of values of h for N rows deep to that for 10 rows deep $\left[\dfrac{h_N}{h_{10}}\right]$

N	1	2	3	4	5	6	7	8	9	10
Staggered tubes	.68	.75	.83	.89	.92	.95	.97	.98	.99	1.0
In-line tubes	.64	.80	.87	.90	.92	.94	.96	.98	.99	1.0

101

NATURAL CONVECTION (Fluid properties are taken at the film temperature, T_f) : $T_f = \frac{1}{2}(T_w + T_\infty)$

Configuration and geometry	Formula and range of validity	Notation
VERTICAL PLATE	$Nu_x = 0.508 \, Pr^{0.5}[0.952 + Pr]^{-0.25} Gr_x^{0.25}$ for $Gr < 5 \times 10^8$ $\dfrac{Nu}{Nu_x} = 1.31 \left[\dfrac{Gr}{Gr_x}\right]^{0.25}$ for liquid metals $Gr_x = \dfrac{g\beta(T_w - T_\infty)x^3}{\nu^2}$	Nu_x, local Nusselt Number Nu, average Nusselt Number Pr, Prandtl Number Gr_x, local Grashof Number Gr, Grashof Number at $x = \ell$ where ℓ is the height of the plate. g, gravitational constant β, coefficient of cubical expansion T_w, wall temperature T_∞, free stream fluid temperature ν, kinematic viscosity
VERTICAL SURFACE	$Nu = C[GrPr]^n$ where 	C	n	GrPr
refer chart alongside		0.1 to 10^4
0.59	0.25	10^4 to 10^9
0.13	0.33	10^9 to 10^{12}		
HORIZONTAL CYLINDERS	$Nu = C[GrPr]^n$ where 	C	n	GrPr
0.4	0	0 to 10^{-5}
Refer chart alongside		10^{-5} to 10^4
0.53	0.25	10^4 to 10^9
0.13	0.33	10^9 to 10^{12}		

Chart (Vertical plates): y-axis $\log(Nu_f)$ with values 2.6, 2.2, 1.8, 1.4, 1.0, 0.6, +0.2, -0.2; x-axis $\log(Gr_f Pr_f)$ with values -1, +1, 3, 5, 7, 9, 11.

Chart (Horizontal cylinders): y-axis $\log(Nu_f)$ with values 2.2, 1.8, 1.4, 1.0, 0.6, +0.2, -0.2, -0.6; x-axis $\log(Gr_f Pr_f)$ with values -5, -3, -1, +1, 3, 5, 7, 9.

NATURAL CONVECTION (Cont.)

Configuration and geometry	Formula	Range of validity
HORIZONTAL SQUARE PLATES: i) upper surface of heated plates or lower surface of cooled plates	$\begin{cases} Nu = 0.54(GrPr)^{0.25} \\ Nu = 0.14(GrPr)^{0.33} \end{cases}$	$\begin{cases} 10^5 < GrPr < 2 \times 10^7 \\ 2 \times 10^7 < GrPr < 3 \times 10^{10} \end{cases}$
ii) lower surface of heated plates or upper surface of cooled plates	$Nu = 0.27(GrPr)^{0.25}$	$3 \times 10^5 < GrPr < 3 \times 10^{10}$
iii) for horizontal rectangular plates [for the conditions (i) and (ii) above use the corresponding formulae calculating Gr and Nu on the basis of the longer side of the rectangle]	—	—

SIMPLIFIED EXPRESSIONS FOR AIR

Geometry	h, kcal/m²hrK	h, W/m²K
I) Horizontal plate, heated surface facing up	$2.14(\Delta T)^{0.25}$	$2.49(\Delta T)^{0.25}$
ii) Horizontal plate, heated surface facing down	$1.13(\Delta T)^{0.25}$	$1.31(\Delta T)^{0.25}$
iii) Vertical plate for height H > 0.3 m	$1.52(\Delta T)^{0.25}$	$1.77(\Delta T)^{0.25}$
iv) Vertical plate for height H < 0.3 m	$1.18(\Delta T/H)^{0.25}$	$1.37(\Delta T/H)^{0.25}$
v) Horizontal pipe of diameter, d	$1.13(\Delta T/d)^{0.25}$	$1.31(\Delta T/d)^{0.25}$
vi) Vertical pipe of diameter d and height H > 0.3 m	$1.13(\Delta T/d)^{0.25}$	$1.31(\Delta T/d)^{0.25}$

where $\Delta T = T_w - T_\infty$

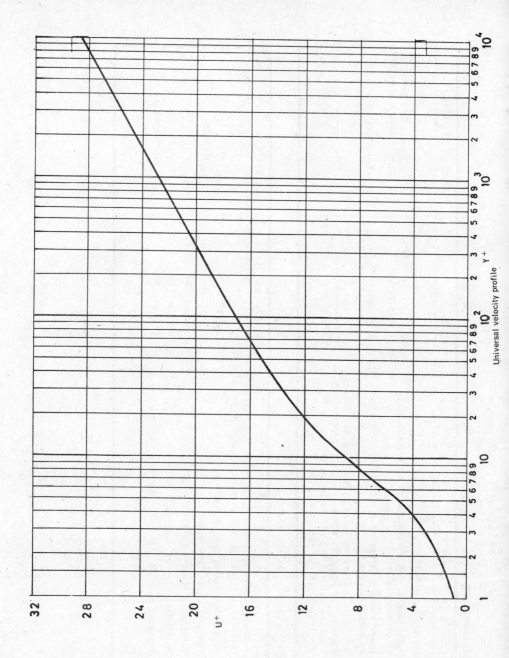

Universal velocity profile

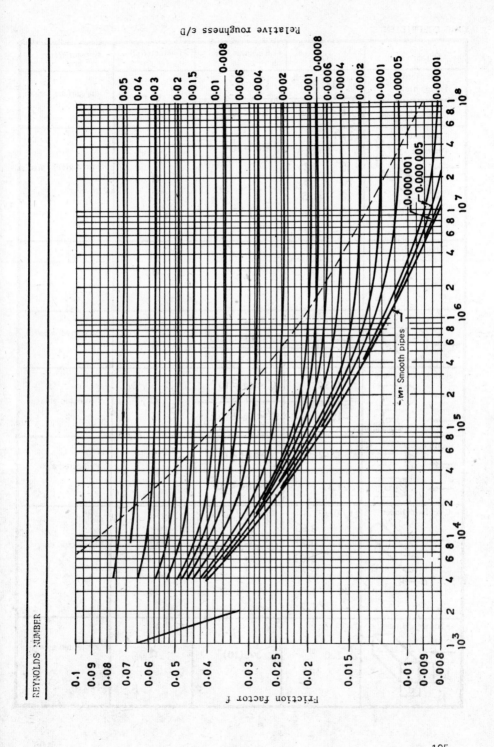

REYNOLDS NUMBER

Friction factor f

Relative roughness e/D

105

DRAG COEFFICIENTS

Object	C_D	Reynolds No. range	Characteristic length	Characteristic area
Flat plate (tangential)	$1.33(Re)^{-.5}$	Laminar	L	Plate surface area
	$0.074(Re)^{-.2}$	$Re < 10^7$		
Flat plate (normal)	L/d 1 1.18 5 1.2 10 1.3 20 1.5 30 1.6 ∞ 1.95	$Re > 10^3$	d	Plate surface area
Circular disc (normal)	1.17	$Re > 10^3$	d	
Sphere	$24(Re)^{-5}$	$Re < 1$	d	Projected area
	0.47	$10^3 < Re < 3 \times 10^5$		
	0.2	$Re > 3 \times 10^5$		
Hollow hemisphere	0.34	$10^4 < Re < 10^6$	d	Projected area
	1.42	$10^4 < Re < 10^6$		
Solid hemisphere	0.42	$10^4 < Re < 10^6$	d	Projected area
	1.17	$10^4 < Re < 10^6$		
Circular cylinder	L/D 1 0.63 5 0.8 10 0.83 20 0.93 30 1.0 ∞ 1.2	$10^3 < Re < 10^5$	d	Projected area
Square cylinder	2.0	$3.5(10)^4$	d	Projected area

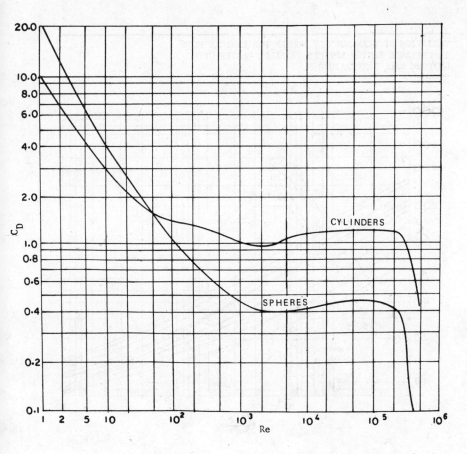

VARIATION OF MEAN NUSSELT NUMBER FOR LAMINAR FLOW IN ENTRANCE REGION BETWEEN PARALLEL PLATES WITH UNIFORM WALL TEMPERATURE

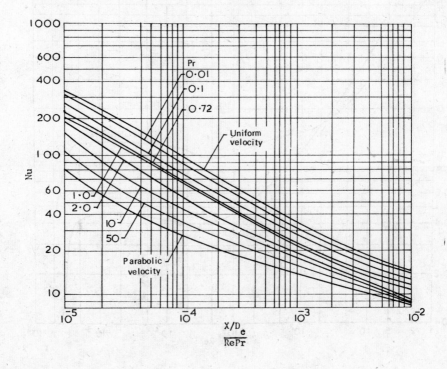

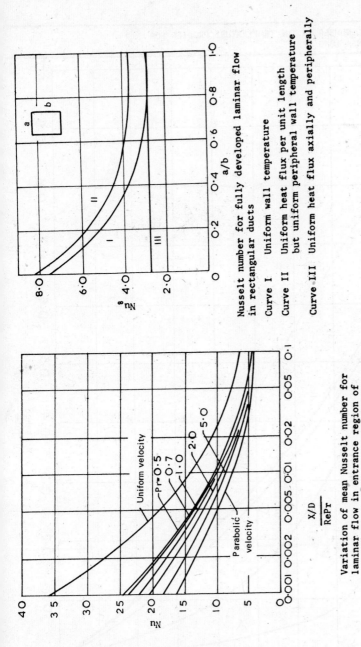

Nusselt number for fully developed laminar flow in rectangular ducts

Curve I   Uniform wall temperature

Curve II   Uniform heat flux per unit length
            but uniform peripheral wall temperature

Curve III   Uniform heat flux axially and peripherally

Variation of mean Nusselt number for laminar flow in entrance region of tubes with uniform wall temperature

EFFECT OF PRANDTL NUMBER ON THE TEMPERATURE PROFILE FOR
TURBULENT FLOW IN A LONG PIPE

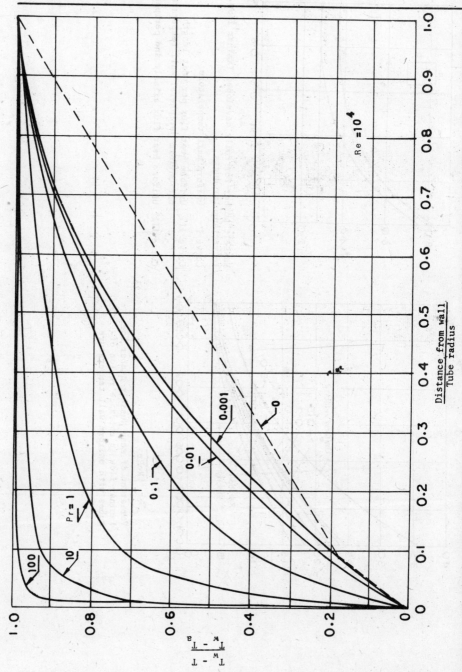

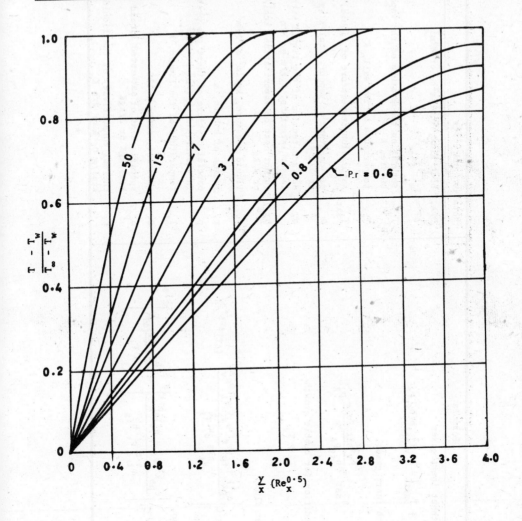

BOILING

Type of boiling	Formula	Notation and units
NUCLEATE	without inert gases $T_v - T_{sat} \approx 2R_v T_{sat}^2 \sigma / h_{fg} p_\ell r$ with inert gases, $T_v - T_{sat} = \dfrac{R_v T_{sat}^2}{h_{fg} p_\ell}\left(\dfrac{2\sigma}{r} - p_g\right)$ Condition for bubble growth $\qquad T_\ell > T_v$ Heat flux $\dfrac{Q}{A} = C\left[\dfrac{c_{p\ell} k^2 h_{fg}^3 (\rho_\ell - \rho_v)\, \rho_\ell^{0.5} \rho_v^3}{\mu_\ell^{0.5}\, \sigma^4\, T_\ell^4}\right]^{0.5} (\Delta T)^{3.5}$ Also, $\dfrac{Q}{A} = C_1\left[\dfrac{c_{p\ell} k^2 (\rho_\ell - \rho_v)\, \rho_\ell^{0.5}}{\mu_\ell^{0.5}\, \rho_v\, h_{fg}}\right]^{0.5} \left[\dfrac{h_{fg}\, \rho_v (\Delta T)}{\sigma\, T_\ell}\right]^{2.3}$	Q/A, heat flux, kcal/m^{2}h or W/m^2 $\Delta T = T_w - T_{sat}$ T_v, saturation temperature of the vapour bubble, K T_w, temperature of the wall, K T_{sat}, saturation temperature of liquid, K R_v, gas constant of vapour, kcal/K kg or J/K kg σ, surface tension of the liquid with vapour interface, kgf/m or N/m h_{fg}, latent heat of vaporisation, kcal/kg or J/kg p_ℓ, liquid pressure, kgf/m^2 or N/m^2 r, radius of bubble, m p_g, partial pressure of inert gas present in the liquid, kgf/m^2 or N/m^2 T_ℓ, temperature of the liquid, K C, constant, refer p.115 $c_{p\ell}$, specific heat of saturated liquid kcal/kgK or J/kgK k, thermal conductivity of the liquid, kcal/m h K or W/m K

BOILING : (Cont.)

Type of boiling	Formula	Notation and units
FILM BOILING	on a horizontal tube : $$h_c = 0.76 \left[\frac{k_v^3 \rho_v (\rho_\ell - \rho_v) g (h_{fg} + 0.4 c_{pv} \Delta T)}{\mu_v d \, \Delta T} \right]^{0.25}$$ $$h_r = \sigma_r \, \epsilon \left[\frac{T_w^4 - T_{sat}^4}{T_w - T_{sat}} \right]$$ $$h = h_c (h_c/h)^{0.33} + h_r$$	ρ_ℓ, density of liquid, kg/m³ ρ_v, density of vapour, kg/m³ μ_ℓ, abs. viscosity of liquid kg/m h or kg/ms C_1, 0.001638 (MKS) = 1.464 × 10⁻⁹ (SI) h_c, heat transfer coefficient due to film conduction, kcal/m²h K or W/m² K h_r, heat transfer coefficient due to radiation k_v, thermal conductivity of vapour, kcal/m h K or W/m K g, acceleration due to gravity m²/h. or m²/s c_{pv}, specific heat of vapour, kcal/kg K or J/kgK μ_v, abs. viscosity of vapour, kg/m h or kg/m s d, diameter of tube, m σ_r, Stefan-Boltzmann constant ϵ, emissivity h, total heat transfer coefficient Pr_ℓ, Prandtl Number of saturated liquid C_{sf}, constant, refer p.115 g_o, 1.27 × 10⁶ in MKS or 1.0 in SI
POOL BOILING	$$c_{p\ell} \Delta T / h_{fg} = C_{sf} \left[\frac{(Q/A)}{\mu_\ell h_{fg}} \frac{g_o}{g} \{ \sigma/(\rho_\ell - \rho_v) \}^{0.5} \right]^{0.33} Pr_\ell^{1.7}$$	

SIMPLIFIED EXPRESSIONS FOR BOILING HEAT TRANSFER COEFFICIENT
FOR WATER AT ONE ATMOSPHERE*

Type of Surface	Range of validity		h		Notation
	kcal/m²h	W/m²	kcal/m²h K	W/m²K	
HORIZONTAL	$\frac{Q}{A} < 12\,500$	$\frac{Q}{A} < 14\,550$	$896.5(\Delta T)^{0.333}$	$1\,042.6(\Delta T)^{0.333}$	$\frac{Q}{A}$, heat flux, kcal/h m² or W/m²
	$12\,500 < \frac{Q}{A} < 187\,500$	$14550 < \frac{Q}{A} < 218\,000$	$4.78(\Delta T)^{3}$	$5.56(\Delta T)^{3}$	$\Delta T = T_w - T_{sat}$
					h_p, heat transfer coefft at pressure p
					p, pressure in number of atmospheres
VERTICAL	$\frac{Q}{A} < 2\,500$	$\frac{Q}{A} < 2\,910$	$462(\Delta T)^{0.143}$	$537.3(\Delta T)^{0.143}$	h, heat transfer coefft at one atm
	$2\,500 < \frac{Q}{A} < 50\,000$	$2\,910 < \frac{Q}{A} < 58\,100$	$6.82(\Delta T)^{3}$	$7.93(\Delta T)^{3}$	

* For other pressures $h_p = h\, p^{0.4}$

114

C_{sf}, CONSTANT FOR POOL BOILING

Surface - fluid combination	C_{sf}
Water - Nickel	0.006
Water - Platinum	0.013
Water - Copper	0.013
Water - Brass	0.006
CCl_4 - Copper	0.013
Benzene - Chromium	0.010
n-Pentane - Chromium	0.015
Ethyl alcohol - Chromium	0.0027
Isopropyl alcohol - Copper	0.0025
35% K_2CO_3 - Copper	0.0054
50% K_2CO_3 - Copper	0.0027
n-Butyl alcohol - Copper	0.0030

C CONSTANT FOR NUCLEATE BOILING

Liquid	Surface	Apparatus		10^6 x C	10^{12} x C
				MKS	SI
Water	Brass	Horizontal	tube	1393.0	7635
	" dirty	"	plate	390.0	2140
	" clean	"	"	89.2	489
	Nickel	"	wire	1393.0	7635
	Chromium	"	tube	222.9	1220
	"	"	plate	8.4	46
Ethanol	Copper	"	"	33.4	183
	Chromium	"	"	8.4	46
n-Butanol	Brass	"	tube	89.2	489
Propane	Copper	"	"	16.7	91.5
	Chromium	"	plate	1.67	9.15
n-Heptane	"	"	"	2.8	15.3
CCl_4	Brass	"	tube	111.5	630
	Chromium	Vertical	"	8.4	46
Methyl chloride	Copper	Horizontal	"	25.1	137.3
Acetone	Chromium	"	plate	19.5	107

LATENT HEAT OF VAPORISATION FOR H_2O

Temp. °C	LATENT HEAT kcal/kg	kJ/kg
0	593.7	2501.6
50	569.0	2382.9
100	539.0	2256.9
150	505.0	2113.2
200	463.5	1938.6
250	409.7	1714.6
300	335.4	1406.0
350	213.3	895.7
374.15	0.0	0.0

SURFACE TENSION, σ, OF WATER AGAINST AIR

Temp. °C	$\sigma \times 10^6$ kgf/cm	$\sigma \times 10^3$ N/m
0	79.1	77.6
10	75.7	74.2
20	74.3	72.9
30	72.5	71.1
40	71.1	69.7
50	69.2	67.9
60	67.5	66.2
70	65.7	64.4
80	63.9	62.7
100	60.0	58.8

LATENT HEAT OF VAPORIZATION

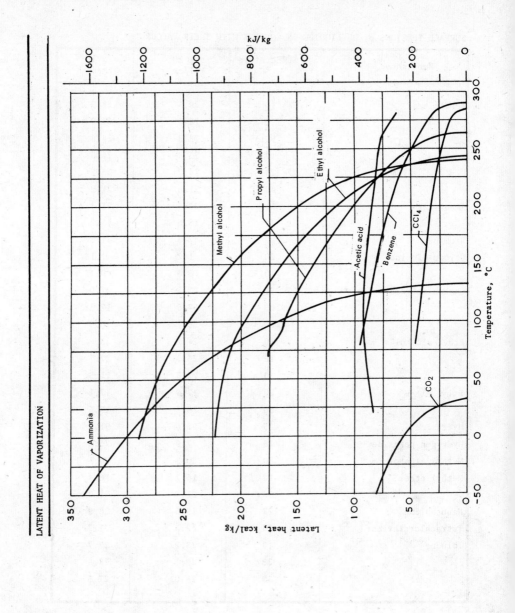

SURFACE TENSION, σ, OF LIQUIDS IN CONTACT WITH THEIR VAPOUR

Substance	Temperature °C	$\sigma \times 10^6$ kgf/cm	$\sigma \times 10^3$ N/m
Acetic acid	10	29.5	28.9
	20	28.4	27.9
	50	25.3	24.8
Acetone	0	26.8	26.3
	20	24.2	23.7
	40	21.6	21.2
Ammonia	11	23.9	23.4
	34	18.6	18.2
Bromine	20	42.3	41.5
Carbon dioxide	20	1.2	1.18
	-10.55	9.3	9.12
Carbon tetrachloride	20	27.5	27.0
	100	17.6	17.3
	200	6.7	6.6
Ethyl alcohol	10	24.1	23.6
	20	23.2	22.7
	30	22.3	21.9
Ethyl ether	20	17.4	17.1
	50	13.7	13.4
Hydrazine	25	93.6	91.8
Hydrogen peroxide	18	77.9	76.4
Methyl alcohol	50	20.5	20.1
Methyl ether	-10	16.7	16.4
	-40	21.4	21.0
Naphthalene	127	29.5	28.9
Tetrachlorethylene	20	32.3	31.7
Toluene	10	28.2	27.7
	20	29.1	28.5
	30	28.0	27.5

CRITICAL STATE CONDITIONS

Material	Temp. T_c °C	abs. pressure P_c kgf/cm^2	abs. pressure P_c N/mm^2	Sp. Vol V_c x 10^3 m^3/kg
Air	-140.70	38.4	3.77	3.22
Methyl alcohol	240.00	81.3	7.97	3.68
Ethyl alcohol	243.33	65.1	6.38	3.62
Ammonia	132.3	115.0	11.28	4.26
Argon	-122.00	49.6	4.86	1.88
Butane	152.5	37.1	3.64	4.43
Carbon dioxide	31.1	75.4	7.39	2.14
Carbon monoxide	-140.00	35.6	3.49	3.32
Carbon tetrachloride	282.75	46.4	4.55	1.81
Chlorine	143.85	78.7	7.72	1.74
Ethane	32.20	50.4	4.94	4.74
Ethylene	9.40	59.6	5.84	4.55
Helium	-267.75	23.2	2.28	14.41
Hexane	235.00	30.5	2.99	4.24
Hydrogen	-240.00	13.2	1.29	32.19
Methane	-82.20	47.3	4.64	6.18
Methyl Chloride	143.30	67.9	6.66	2.68
Neon	-228.85	27.4	2.69	1.05
Nitric oxide	-93.85	67.2	6.59	1.93
Nitrogen	-147.20	34.5	3.38	3.24
Octane	296.10	25.4	2.49	4.24
Oxygen	-118.85	51.3	5.03	2.31
Propane	95.55	44.4	4.35	4.43
Sulphur dioxide	157.20	80.2	7.86	1.93
Water	374.15	225.7	22.12	3.26
Refrigerant 12	111.7	40.8	4.00	1.79
Refrigerant 11	198.0	44.6	4.37	1.81
Refrigerant 13	28.8	39.4	3.87	1.72

CONDENSATION

$Q = hA \Delta T$ [Properties of Fluid are to be taken at T_f]

Geometry and conditions	Formula	Notation
FILM CONDENSATION: i) VERTICAL SURFACES (laminar) ii) VERTICAL TUBES iii) HORIZONTAL TUBES iv) BANK OF TUBES	$\delta_x = \left[\dfrac{4\mu k \, x \, (T_f - T_s)}{g \, h_{fg} \, \rho^2} \right]^{0\cdot25}$ $h_x = \dfrac{k}{\delta_x}$ $h = \dfrac{4}{3} h_x$ (at x = 1) $h = 0.943 \left[\dfrac{k^3 \, \rho^2 \, g \, h_{fg}}{\mu \, \ell \, (T_f - T_s)} \right]^{0\cdot25}$ $h = 0.728 \left[\dfrac{k^3 \, \rho^2 \, g \, h_{fg}}{\mu \, d \, (T_f - T_s)} \right]^{0\cdot25}$ $h = 0.728 \left[\dfrac{k^3 \rho^2 g h_{fg}}{\mu n d (T_f - T_s)} \right]^{0\cdot25}$	δ_x, Boundary layer thickness, m μ, Abs. viscosity of fluid, kg/m h: or kg/ms k, Thermal conductivity of the fluid, kcal/m h °K or W/mK x, Distance along the surface, m T_f, Film temperature, K $\quad = \dfrac{T_\infty + T_s}{2}$ T_s, Surface temperature, K T_∞, Free stream temperature, K g, Gravitational constant m/h ^{2} or m/s^2 h_{fg}, Latent heat of vaporisation kcal/kg or J/kg ρ, Density of fluid, kg/m^3 h_x, Local heat transfer coefft. } kcal/m^{2}h or h, Average heat transfer coefft. } W/m^2 K

120

CONDENSATION (Cont.)

Geometry and Conditions	Formula	Notations
v). VERTICAL TUBES	$h = \dfrac{1.47}{Re^{0.33}}(k^3\rho^2 g/\mu^2)^{0.33}$ for Re < 1800 $h = 0.0077 Re^{0.4}(k^3\rho^2 g/\mu^2)^{0.33}$ for Re > 1800	ℓ, Lenght of tube, m d, Diameter of tube, m n, Number of horizontal rows placed one above the other (irrespective of the number of tubes in a horizontal row) Re, Reynolds Number $= \dfrac{4\,A\,\rho\,u}{P\,\mu}$ A, flow area, m^2 P, shear perimeter, m u, average velocity of flow Z, condensation coefft. (Refer table on page 122) W, weight of condensate in kgf/h or N/h N_v, number of vertical columns ℓ, length of tube, cm N, total number of tubes d, diameter (outer) of tubes, cm
vi) VERTICAL SHEET CONTAINING HORIZONTAL TUBES	$h = \dfrac{1.51}{Re^{0.33}}\left(\dfrac{k^3\,\rho^2 g}{\mu^2}\right)^{0.33}$ for Re < 1800	
S I M P L I F I E D E X P R E S S I O N S		
i) Single vertical tube	$h \doteq 1.26\,Z\left[\dfrac{\pi\,d}{W}\right]^{0.33}$	
ii) a bank of horizontal tubes	$h = Z\left[\dfrac{N_v\,\ell}{W}\right]^{0.33}$	
iii) Vertical Condensers	$h = 1.26\,Z\left[\dfrac{N\,\pi\,d}{W}\right]^{0.33}$	

CONDENSATION COEFFICIENT FOR FLUIDS

Fluid	† Condensation coefficient Z					
	0°C	50°C	100°C	150°C	200°C	300°C
Acetone	1255	1291	1278			
Ammonia	3960					
Benzene	968	1125	1176	1280		
Deccine		772	820	828	772	
Dodeccine			790	811	811	717
Dowtherm-A					963	1000
Ethyl alcohol	772	963	1128			
Ethyl glycol		852	1060	1150		
Gasoline		820	845	845		
Hexane	866	900	907	875		
Kerosene			797	820	802	
Methyl alcohol	1150	1361	1568			
Natural gasoline	852	930	975			
Octane	797	845	882	859		
Pentane	937	937	907	890		
i-Propyl alcohol		715	981			
Steam		4560	6190	7750		
Tetradeccine			742	780	790	731
Toluene		1071	1149			
Xylene		1155	1109			

† For use with SI units multiply the above values of Z by 2.49

HEAT EXCHANGERS $\quad Q = UA\,(\Delta T)_{\ell m}$

Type	Formula		
SINGLE PASS i) Parallel flow	$(\Delta T)_{\ell m} = \dfrac{(T_1 - t_1) - (T_2 - t_2)}{\ln\left[\dfrac{T_1 - t_1}{T_2 - t_2}\right]}$	$Q,$	Heat exchanged, kcal/h. or W
		$U,$	Overall heat transfer coefft. W/m²K or kcal/m² h °K (Refer pages 124 & 125)
		$(\Delta T)_{\ell m},$	Logarithmic mean temperature difference, "LMTD"
ii) Counter flow	$(\Delta T)^*_{\ell m} = \dfrac{(T_1 - t_2) - (T_2 - t_1)}{\ln\left[\dfrac{(T_1 - t_2)}{(T_2 - t_1)}\right]}$	$A,$	area, m²
		$T_1,$	Entry temperature of hot fluid
		$T_2,$	Exit temperature of hot fluid
		$t_1,$	Entry temperature of cold fluid
		$t_2,$	Exit temperature of cold fluid
MULTIPASS and CROSS FLOW	$Q = F U A\,(\Delta T)^*_{\ell m}$	$F,$	CORRECTION FACTOR depending on R, P and type of exchanger (Refer pages 124 to 127)
NTU METHOD	$Q = \epsilon C_{min}(T_1 - t_1)$	$P = \dfrac{t_2 - t_1}{T_1 - t_1}\;;\quad R = \dfrac{T_1 - T_2}{t_2 - t_1}$	
		$NTU,$	Number of Transfer Units $= AU/C_{min}$
	Note: $\dfrac{C_{min}}{C_{max}}$ is zero when one of the fluids is condensing or evaporating.]	$C_{min},$	Smaller value of $m_h c_h$ and $m_c c_c$; kcal/h °K or W/K
		$m_h,$	Mass flow rate of hot fluid, kg/h or kg/s
	$\epsilon = \dfrac{m_h c_h}{C_{min}}\left[\dfrac{T_1 - T_2}{T_1 - t_1}\right]$	$m_c,$	Mass flow rate of cold fluid
		$c_h,$	Specific heat of hot fluid, kcal/kg K or J/kg K
	$= \dfrac{m_c c_c}{C_{min}}\left[\dfrac{t_2 - t_1}{T_1 - t_1}\right]$	$c_c,$	Specific heat of cold fluid
		$\epsilon,$	Effectiveness (depends on C_{min}/C_{max}, NTU and geometry; refer charts on pages 128 to 133)

APPROXIMATE OVER-ALL COEFFICIENTS FOR PRELIMINARY ESTIMATES

Duty	Overall coefficient, U	
	kcal/h m^2 °K	W/m^2 K
Steam to water-instantaneous heater	2000-3000	2326-3489
Steam to water-storage-tank heater	850-1500	989-1745
Steam to heavy fuel oil	50- 150	58- 174
Steam to light fuel oil	150- 300	174- 349
Steam to light petroleum distillate	250-1000	291-1163
Steam to aqueous solutions	500-3000	582-3489
Steam to gases	25- 250	29- 290
Water to compressed air	50- 150	58- 174
Water to water, jacket water coolers	750-1350	872-1570
Water to lubricating oil	100- 300	116- 349
Water to condensing oil vapours	200- 500	233- 582
Water to condensing alcohol	225- 600	262- 698
Water to condensing Refrigerant-12	400- 750	465- 872
Water to condensing ammonia	750-1250	872-1454
Water to organic solvents, alcohol	250- 750	291- 872
Water to boiling Refrigerant-12	250- 750	291- 872
Water to gasoline	300- 450	349- 523
Water to gas oil for distillate	175- 300	204- 349
Water to brine	500-1000	582-1163
Light organics to light organics	200- 375	233- 436
Medium organics to medium organics	100- 300	116- 349
Heavy organics to heavy organics	50- 200	58- 233
Heavy organics to light organics	50- 300	58- 349
Crude oil to gas oil	150- 275	174- 320

FOULING RESISTANCE, R

Type of Fluid	Fouling resistance	
	h °K m^2/kcal	Km^2/W
Sea water below 52°C	0.000102	0.0000877
Sea water above 52°C	0.000204	0.0001754
Treated boiler feed water above 52°C	0.000204	0.0001754
Fuel oil	0.00102	0.0000877
Quenching oil	0.00082	0.0007051
Alcohol vapours	0.000102	0.0000877
Steam	0.000102	0.0000877
Industrial air	0.0041	0.003525
Refrigerant	0.00204	0.001754

Dirt film which gradually builds up on the heat transfer surface increases the thermal resistance; and this is accounted for by FOULING FACTORS, R_o and R_i of the outer and inner surfaces of heat transfer, viz A_o and A_i.

$$\frac{1}{U_o} = \frac{1}{h_o} + R_o + R_{ko} + \frac{R_i A_o}{A_i} + \frac{A_o}{A_i h_i}$$

where R_{ko} is wall resistance to heat conduction and U_o is the overall heat transfer coefficient, both based on the outer area.

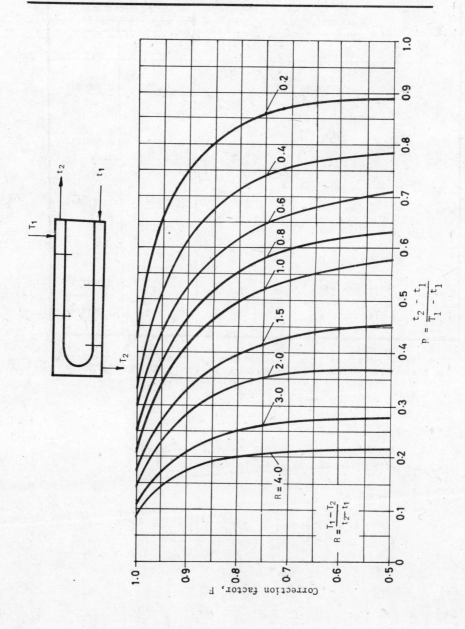

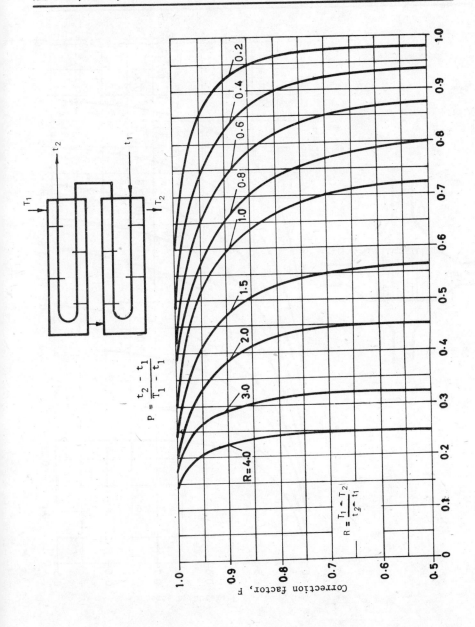

CORRECTION FACTOR PLOT FOR SINGLE PASS CROSS-FLOW EXCHANGER
(ONE FLUID MIXED; OTHER UNMIXED)

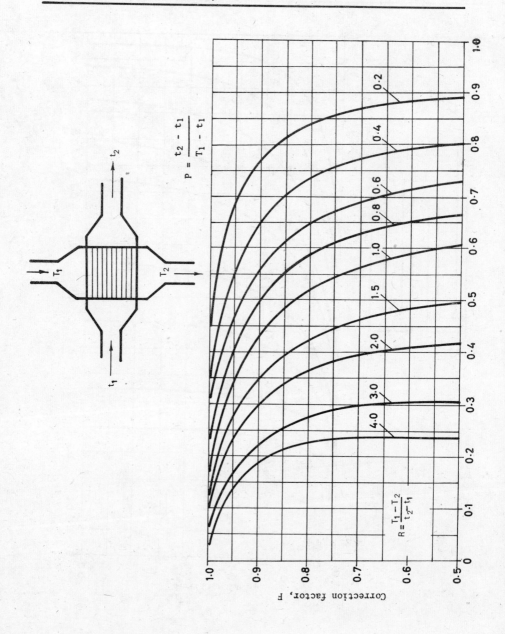

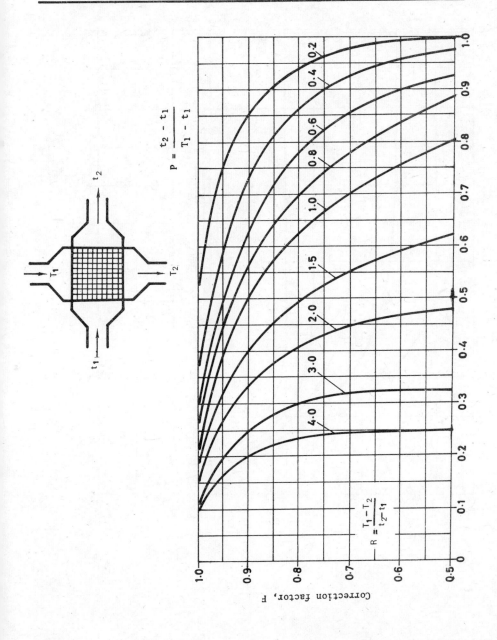

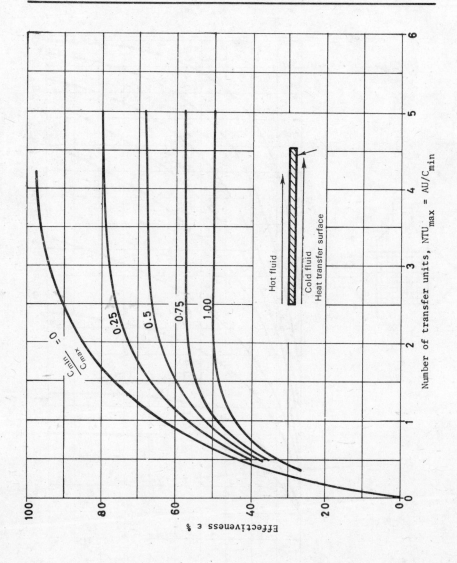

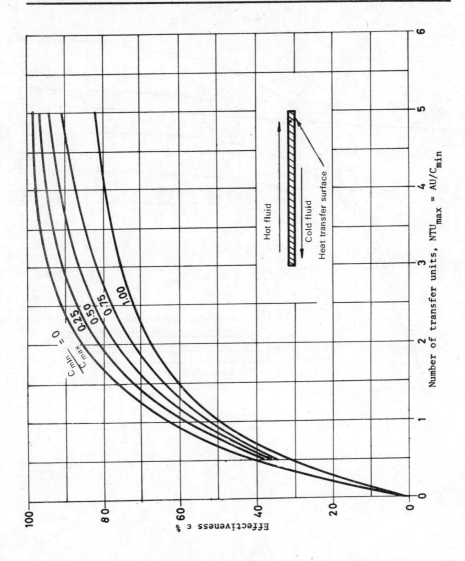

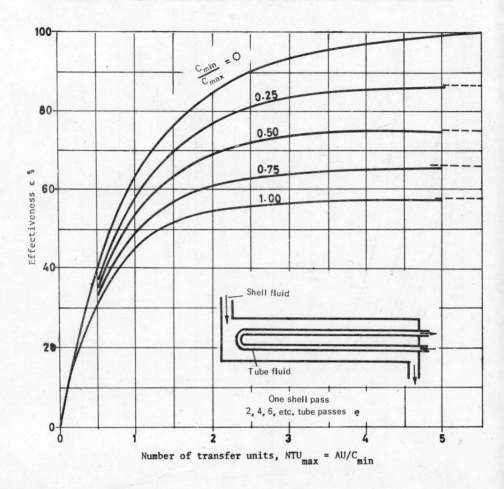

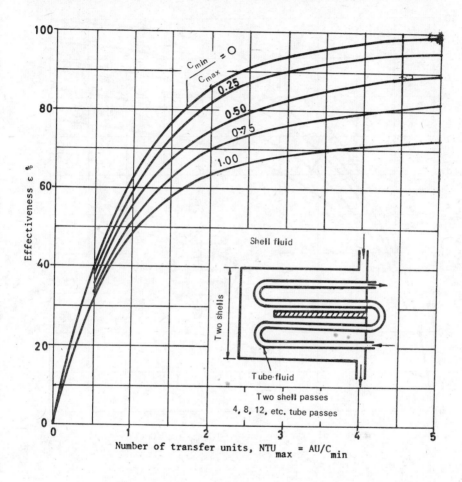

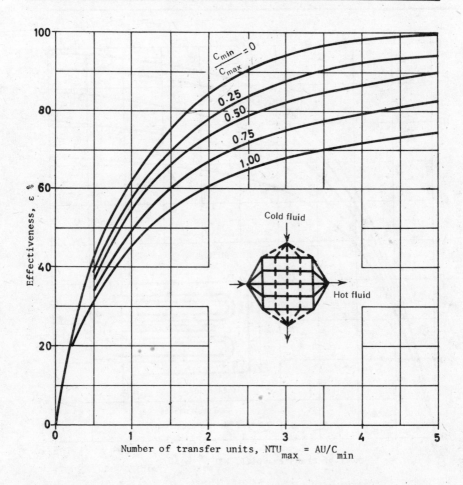

EFFECTIVENESS FOR CROSS FLOW EXCHANGER WITH ONE FLUID MIXED

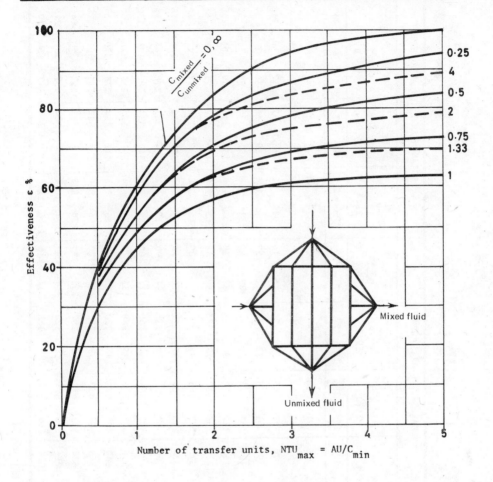

TUBE SHEET LAYOUT, number of tubes

1P = single pass; 2P = two pass

SHELL DIAMETER mm	SQUARE PITCH ARRANGEMENT								TRIANGULAR PITCH ARRANGEMENT							
TUBE OD, mm / pitch, mm	19 / 25.4		25.4 / 31.0		31.0 / 39.7		38.1 / 47.6		19 / 25.4		25.4 / 31.0		31.0 / 39.7		38.1 / 47.6	
	1P	2P	1P	2P	1P	2P	1P	2P	1P	2P	1P	2P	1P	2P	1P	2P
203.2	32	26	21	16	--	--	--	--	37	30	21	16	--	--	--	--
254.0	52	52	32	32	16	12	--	--	61	52	32	32	20	18	--	--
304.8	81	76	48	45	30	24	16	16	92	82	55	52	32	30	18	14
336.6	97	90	61	56	32	30	22	22	109	106	68	66	38	36	27	22
387.4	137	124	81	76	44	40	29	29	151	138	91	86	54	51	36	34
438.2	177	166	112	112	56	53	39	39	203	196	131	118	69	66	48	44
489.0	224	220	138	132	78	73	50	48	262	250	163	152	95	91	61	58
539.8	277	270	177	166	96	90	62	60	316	302	199	188	117	112	76	72
590.6	341	324	213	208	127	112	78	74	384	376	241	232	140	136	95	91
635.0	413	394	260	252	140	135	94	90	470	452	294	282	170	164	115	110
685.8	481	460	300	288	166	160	112	108	559	534	349	334	202	196	136	131
736.6	553	526	341	326	193	188	131	127	630	604	397	376	235	228	160	154
787.4	657	640	406	398	226	220	151	146	745	728	472	454	275	270	184	177
838.2	749	718	465	460	258	252	176	170	856	830	538	522	315	305	215	206
889.0	845	824	522	518	293	287	202	196	970	938	608	592	357	348	246	238
939.8	934	914	596	574	334	322	224	220	1074	1044	674	664	407	390	275	268
990.6	1049	1024	665	644	370	362	252	246	1206	1176	766	736	449	436	307	299

136

MASS TRANSFER

MOLECULAR DIFFUSION :

Condition	Equation	Notation
1. Fick's Law of diffusion (of component A into component B) for solids, liquids and gases	$\dfrac{N_a}{A} = -D\left(\dfrac{\partial c_a}{\partial y}\right)$ $\dfrac{\dot{m}_a}{A} = -D\left(\dfrac{\partial c_a}{\partial y}\right)$	$\dfrac{N_a}{A}$, mass flux, kg mole/m²h $\dot{N}_a$, kg mole/h A, area, m² (normal to direction of diffusion, y) D, diffusion coefficient, m²/h C_a, mass concentration of A per unit volume of mixture of components A and B, kg mole/m³ or kg/m³ $\dot{m}_a$, mass flux per unit time, kg/h D_{ab}, diffusion coefficient when A diffuses into B R_a, gas constant of A; $R_a = \dfrac{\text{universal gas const.}}{\text{molecular weight of A}}$ p_a, partial pressure of component A T, absolute temperature of mixture, K $dp_a = p_{a2} - p_{a1}$ $dy = y_2 - y_1$
2. Fick's Law for gases (steady state equi-molal counter diffusion) $N_a = N_b$	$\dfrac{\dot{m}_a}{A} = -D_{ab}\left(\dfrac{1}{R_a T}\dfrac{dp_a}{dy}\right)$ $\dfrac{\dot{m}_b}{A} = -D_{ba}\left(\dfrac{1}{R_a T}\dfrac{dp_b}{dy}\right)$	
3. Steady state diffusion of A into stagnant component B [Stefan's Law]	$\dfrac{\dot{m}_a}{A} = D_{ab}\,\dfrac{p}{R_a T(y_2 - y_1)}\,\ell n(p_{b2}/p_{b1})$ $= D\,\dfrac{p}{R_a T(y_2 - y_1)P_{bm}}\,(p_{b2} - p_{b1})$ _Note:-_ $p_{a1} - p_{a2} = p_{b2} - p_{b1}$	

MASS TRANSFER (Cont.): MOLECULAR DIFFUSION :

Condition	Equation	Notation
4. Steady state diffusion of A into a stagnant mixture of components B, C, D, etc.	$$\frac{\dot{m}_a}{A} = D_{a\,mix}\frac{p}{R_a T(y_2 - y_1)}\ln\frac{p_{mix2}}{p_{mix1}}$$	p_{bm}, logarithmic mean partial pressure difference $= (p_{b2} - p_{b1})/\ln(p_{b2}/p_{b1})$ $D_{a\,mix}$, diffusion coefft. when A diffuses into a mixture of components B, C, D, etc. $$D_{a\,mix} = \frac{1}{\left[\dfrac{n_b}{D_{ab}} + \dfrac{n_c}{D_{ac}} + \dfrac{n_d}{D_{ad}} + \cdots\right]}$$ n_b, mole fraction of component B in mixture before diffusion p, total pressure p_{mix}, partial pressure of the mixture y, = distance along the direction of diffusion $c = c_a + c_b$ c_{bm}, logarithmic mean concentration difference $= (c_{b2} - c_{b1})/\ln(c_{b2}/c_{b1})$
5. Steady state equi-molal counter diffusion between liquids	$$\frac{N_a}{A} = D_{ab}\left[\frac{c_{a1} - c_{a2}}{y_2 - y_1}\right]$$	
6. Steady state diffusion of liquid A into stagnant liquid B ($N_b = 0$)	$$\frac{N_a}{A} = D_{ab}\left[\frac{c_{a1} - c_{a2}}{y_2 - y_1}\right]\frac{c}{c_{bm}}$$	

CONVECTIVE MASS TRANSFER : $\dot{m}_a = KA(c_{a1} - c_{a2})$

K, mass transfer coefft. m/h
c_a, concentration

Conditions	Equation to determine K	Notation
1. Steady state diffusion **across** a thickness Δy	$$K = \frac{D}{\Delta y}$$	For gases $c_a = \dfrac{p_a}{M_a R_a T} = \dfrac{p_a}{m_a}$ M_a, molecular weight of A R_a, gas constant of A m_a, Henry's Number at temp. T K D, diffusion coefft. from table on pages 144 to 145
2. Steady state diffusion of A into stagnant component B (for gases)	$$K = \frac{D_{ab}\, p}{(y_2 - y_1)(p_{a1} - p_{a2})} \ln \frac{p_{b2}}{p_{b1}}$$	P, total pressure p_a, p_b, partial pressures Sh, Sherwood Number = $\dfrac{Kx}{D}$ x, a characteristic dimension, m
3. Mass transfer of B into a turbulent flow stream of A in a closed circular interface (A does not diffuse into B)	$Sh = 0.023\ Re^{0.83}\ Sc^{0.44}$ for $2000 < Re < 35000$ and $0.6 < Sc < 2.5$	For circular section, x = diameter Sc, Schmidt Number = $\dfrac{\nu}{D}$ ν, kinematic viscosity
4. For rough pipes. (Reynolds analogy)	$$\frac{K}{u}\, Sc^{0.67} = \frac{f}{8}$$	f, friction factor for rough pipe (Refer chart on page 105) u, average velocity of the flowing fluid, m/h

CONVECTIVE MASS TRANSFER : (Cont.)

Conditions	Equation to determine K	Notation
5. Flow over smooth flat plates		C_f, coefficient of friction
Laminar	$\dfrac{K}{u_\infty} Sc^{0.67} = \dfrac{C_f}{2} = \dfrac{0.332}{Re^{0.5}}$	
Turbulent	$\dfrac{K}{u_\infty} Sc^{0.67} = \dfrac{C_f}{2} = \dfrac{0.0288}{Re^{0.2}}$	u_∞, free stream velocity
6. Simultaneous convective heat and mass transfer	$\dfrac{h}{K} = \rho\, C_p\, Le^{0.67}$	h, heat transfer coefficient, kcal/m² hr K or W/m²K
		ρ, density, kg/m³
		C_p, specific heat at constant pressure, kcal/kg K or J/kg K
		Le, Lewis Number $= \dfrac{Sc}{Pr} = \dfrac{\alpha}{D}$
		α, thermal diffusivity, m²/h
7. Mass transfer through an interface	$\dfrac{1}{K} = \dfrac{1}{K_1} + \dfrac{1}{K_2}$	K_1, mass transfer coefficient from A to interface
		K_2, mass transfer coefficient from interface to medium B
8. Packed beds	Refer packed beds under "CONVECTION"	

HUMIDIFICATION

Conditions	Equation	Notation
1. Water and air in equilibrium — air at constant temperature.	$$p_w - p_g = \frac{h}{h_{fg} K_g}(T_g - T_w)$$	p_w, partial pressure of water vapour at wet bulb temperature, T_w
		p_g, partial pressure of water vapour in air
		h, convective heat transfer coefft. for air to water
		h_{fg}, enthalpy of evaporation for water at temperature, T_w
		K_g, mass transfer coefficient defined through pressure difference $$K_g = \frac{\dot{m}}{A\Delta p}$$
2. — do —	$$Y_w - Y_g = \frac{h}{h_{fg} K_g P}(T_g - T_w)$$ $$Y = \frac{M_w p_w}{M_g P}$$	M_w, molecular weight of water, 18
		Y, abs. humidity of air
		Y_w, abs. humidity of air at temp. T_w (see psychrometric charts)
		Y_g, abs. humidity of air at temp. T_g
		M_g, molecular weight of gas (air)
		P, total pressure
		T_g, temperature of air (dry bulb)

141

HUMIDIFICATION (Cont.)

Conditions	Equation	Notation
3. Water and air in equilibrium Water at constant temperature	$Y_a - Y_g = \dfrac{1}{h_{fg}} [C_{pa} + Y_g C_{pw}] (T_g - T_a)$	Y_a', abs. humidity of air at the final equilibrium air temperature T_a Y_g', abs. humidity of the initial air at the initial air temperature T_g C_{pa}, specific heat of air C_{pw}, specific heat of water
4. Adiabatic humidification	$Z = \dfrac{G}{K_y a} \ln[(Y_a - Y_1)/(Y_a - Y_2)]$ $K_y = \dfrac{\dot{m}}{A (Y_a - Y)}$	Z, height of tower G, mass velocity of air per unit area K_y, gas phase mass transfer coefficient a, interfacial area per unit volume of tower packing Y, abs. humidity of air Y_a, saturated humidity of air at the interface temperature T_a Y_1, humidity of air at entry to tower Y_2, humidity of air at exit of tower A, total area for mass transfer $= a s Z$ $\dot{m}$, mass flux per unit time s, cross sectional area of tower

HUMIDIFICATION (Cont)

Conditions	Equation	Notation
5. Dehumidification	$h_\ell(T_i - T_\ell) = h_g(T_g - T_i)$ $+ h_{fg} K_y (Y_g - Y_i)$	h_ℓ, liquid phase heat transfer coefficient h_g, gas phase heat transfer coefficient T_i, temperature at the interface T_g, bulk gas temperature (dry bulb) T_ℓ, liquid layer temperature Y_i, abs. humidity at temperature T_i

143

DIFFUSION COEFFICIENTS :

Diffusing material (Solute)	Medium of Diffusion (Solvent)	Temperature °C	Diffusion Coefficient $D \times 10^6$ m²/h	Concentration of Solute, c kg mole/m³
Copper	Aluminium	462	3.05	
Mercury	Lead	177	0.704	
Mercury	Lead	197	1.8	
Chlorine	Water	16	4.54	0.12
Hydro chloric acid	Water	0	9.73	9
-do-	-do-	10	6.48	2
-do-	-do-	10	11.9	9
-do-	-do-	16	9	2.5
-do-	-do-	5	8.79	0.5
Ammonia	Water	15	4.46	3.5
-do-	-do-	10	6.37	1
Carbon dioxide	Water	20	5.26	≈ 0
-do-	-do-	18	6.37	≈ 0
Sodium chloride	Water	18	4.54	0.05
-do-	-do-	18	4.35	0.2
-do-	-do-	18	4.46	1.0
-do-	-do-	18	4.9	3.0
-do-	-do-	18	5.55	5.4
Methanol	Water	15	4.61	≈ 0
Acetic acid	Water	12.5	2.96	1
-do-	-do-	12.5	3.28	0.01
-do-	-do-	18	3.46	1
Ethanol	Water	10	1.8	3.75
-do-	-do-	10	2.99	0.05
-do-	-do-	16	3.24	2
n-Butanol	Water	15	2.78	≈ 0
Carbon dioxide	Ethanol	17	11.5	≈ 0
Chloroform	Ethanol	20	4.5	2

DIFFUSION COEFFICIENTS (Cont.)

Solute	Solvent	Temperature °C	$D \times 10^3$ m²/h	Schmidt number, Sc
Ammonia	Air	0	77.8	0.634
Carbon dioxide	Air	0	42.8	1.14
-do-	Hydrogen	18	218	0.158
-do-	Oxygen	0	66.5	
Mercury	Nitrogen	19	11700	0.00424
Oxygen	Air	0	55	0.895
Oxygen	Nitrogen	12	72.6	0.681
Hydrogen	Air	0	197	0.25
Hydrogen	Oxygen	14	279	0.182
Hydrogen	Nitrogen	12.5	266	0.187
Hydrogen	Methane		225	
Water	Air	8	74.1	0.615
-do-	-do-	16	101	0.488
-do-	-do-	26	93	
-do-	-do-	59	110	
Benzene	Air	0	27	1.83
-do-	Carbon dioxide	0	18.95	1.37
-do-	Hydrogen	0	106	3.26
Carbon disulphide	Air	20	31.7	1.68
Ether	Air	20	27.7	1.93
Ethyl Alcohol	Air	0	36.4	1.36
-do-	-do-	40.5	42.5	1.45
	Ethanol	0	36.7	
Air	n-Butanol	26	31.3	
-do-	-do-	59	37.4	
Air	Ethyl acetate	26	31.3	
-do-	-do-	59	38.2	
Air	Aniline	26	26.6	
-do-	-do-	59	32.4	
Air	Chloro benzene	26	26.6	
-do-	-do-	59	32.4	
Air	Toluene	26	31.0	
-do-	-do-	59	32.8	
Carbon monoxide	Oxygen	0	66.6	

SCHMIDT NUMBERS OF VARIOUS SUBSTANCES at 20°C, Sc^*

when in dilute solution in water, ethyl alcohol or benzene. †

Solute	Solvent	Sc^*
Hydrogen	Water	196
Hydrochloric acid	"	381
Nitric acid	"	390
Oxygen	"	558
Carbon dioxide	"	559
Ammonia	"	570
Sulphuric acid	"	580
Nitrogen	"	613
Acetylene	"	645
Nitrous oxide	"	665
Sodium hydroxide	"	665
Hydrogen sulphide	"	712
Sodium chloride	"	745
Methyl alcohol	"	785
Chlorine	"	824
Bromine	"	840
Urea	"	946
Ethyl alcohol	"	1005
Allyl alcohol	"	1080
Urethane	"	1090
Acetic acid	"	1140
Propyl alcohol	"	1150
Phenol	"	1200
Resorcinol	"	1260
Hydroquinone	"	1300
Butyl alcohol	"	1310
Glycerol	"	1400
Pyrogallol	"	1440
Mannitol	"	1730
Sucrose	"	2230
Lactose	"	2340
Raffinose	"	2720
Carbon dioxide	Ethyl alcohol	445
Chloroform	"	1230
Phenol	"	1900
Ethylene dichloride	Benzene	301
Chloroform	"	350
Acetic acid	"	384
Phenol	"	479

† Schmidt Number at any other temperature (Sc) may be found from the following relationship:

$$\frac{Sc}{Sc^*} = \left(\frac{\mu}{\mu^*}\right)^2 \left(\frac{\rho^*}{\rho}\right) \left(\frac{T^*}{T}\right)$$

where

μ, absolute viscosity at absolute temperature T
ρ, density at absolute temperature T
ρ^*, density at 20°C
μ^*, absolute viscosity at 20°C
$T^* = 293$ K

SCHMIDT NUMBERS OF GASES in dilute mixture with air†

GAS	Molecular Weight M	Schmidt Number Sc^*
Hydrogen	2.016	0.22
Methane	16.04	0.84
Ammonia	17.03	0.61
Steam	18.016	0.60
Nitrogen	28.02	0.98
Ethane	30.07	1.22
Oxygen	32.00	0.74
Methyl alcohol	32.04	1.00
Carbon dioxide	44.01	0.96
Propane	44.09	1.51
Ethyl alcohol	46.07	1.30
Acetone	58.08	1.60
Butane	58.12	1.77
Acetic acid	60.05	1.24
n-Propyl alcohol	60.09	1.55
Sulphur dioxide	64.06	1.28
Chlorine	70.90	1.42
Pentane	72.15	1.97
Methyl acetate	74.08	1.57
Ethyl ether	74.12	1.70
n-Butyl alcohol	74.12	1.88
Carbon disulphide	76.13	1.48
Benzene	78.11	1.71
Ethyl acetate	88.10	1.84
Toluene	92.13	1.86
Phosgene	98.92	1.65
n-Propyl acetate	102.13	1.97
Chlorobenzene	112.56	2.13
n-Octane	114.22	2.62
Napthalene	128.16	2.57
Carbon tetrachloride	153.84	2.13
Bromobenzol	157.02	1.97
Chloropicrin	164.39	2.13
Ethylene bromide	187.88	1.97

† Schmidt Numbers of gases in dilute mixture with air may be deduced from the following approximate relationship :

$$Sc^* \simeq 0.145 \, M^{0.556}$$

Schmidt number of a gas mixed with air in any proportion may be determined in the following way :

$$Sc = \frac{\nu}{\nu^*} Sc^*$$

where

ν, viscosity of the mixture, m^2/h

ν^*, viscosity of the component in a dilute mixture at the mixture temperature, m^2/h.

147

BASIC SI UNITS

Physical quantity	Name of unit	Symbol for unit
length	metre	m
mass	kilogram	kg
time	second	s
electric current	ampere	A
temperature	degree Kelvin	K
plane angle	radian	rad
solid angle	steradian	sr

DERIVED SI UNITS

Physical quantity	Name of unit	Symbol for unit	Definition of unit
energy	joule	J	$kg\ m^2\ s^{-2}$
force	newton	N	$kg\ m\ s^{-2}$ or $J\ m^{-1}$
power	watt	W	$kg\ m^2 s^{-3}$ or $J\ s^{-1}$
frequency	hertz	Hz	cycle per second
temperature	degree Celsius	°C	K - 273
time	hour	h	3600 s

CONVERSION FROM MKS TO SI UNITS

MKS	Physical quantity	SI	FPS
L KGF	Force	9.81 N	2.205 lbf
1 kgf/cm^2	Pressure	9.81×10^4 N/m^2	14.22 lbf/in^2
1 atm	-do-	10.13×10^4 N/m^2	14.7 lbf/in^2
1 mm Hg	-do-	133.32 N/m^2	0.01934 lbf/in^2
10 poise	Viscosity	1 N s/m^2	2.089×10^{-3} lbf^{-3}ft^2
1 kgf m	Energy or work	9.81 J	0.1383 ft lbf
1 cal	Heat energy	4.1868 J	3.968×10^{-3} BTU
1 kW	Power	3.6×10^6 J	1.34 hp
1 kgf m/s	-do-	9.81 W	0.1383 ft lbf/s
1 hp	-do-	735.5 W	0.987 hp
1 kcal/hr	Heat flow rate	1.163 W	0.252 BTU/hr
1 kcal/m^2	Heat flow per unit area	1.163 W h/m^2	0.02712 BTU/ft^{2}hr °F
1 kcal/m^2 hr °C	Heat transfer coefficient	1.163 W/m^2 K	0.2048 BTU/ft^{2}hr °F
1 kcal/kg	Enthalpy	4.1868 kJ/kg	0.1145 BTU/lb
1 kcal/kg °C	Specific heat	4.1868 kJ/kg K	0.0635 BTU/lb °F
1 kcal/m hr °C	Thermal conductivity	1.163 W/m K	0.6723 BTU/ft hr °F

A P P E N D I X

BASIC SI UNITS

Physical quantity	Name of unit	Symbol for unit
length	metre	m
mass	kilogram	kg
time	second	s
electric current	ampere	A
temperature	degree Kelvin	K
plane angle	radian	rad
solid angle	steradian	sr

DERIVED SI UNITS

Physical quantity	Name of unit	Symbol for unit	Definition of unit
energy	joule	J	$kg\ m^2\ s^{-2}$
force	newton	N	$kg\ m\ s^{-2}$ or $J\ m^{-1}$
power	watt	W	$kg\ m^2 s^{-3}$ or $J\ s^{-1}$
frequency	hertz	Hz	cycle per second
temperature	degree Celsius	°C	$K - 273$
time	hour	h	3600 s

CONVERSION FROM MKS TO SI UNITS

MKS	Physical quantity	SI	FPS
L KGF	Force	9.81 N	2.205 lbf
1 kgf/cm^2	Pressure	9.81×10^4 N/m^2	14.22 lbf/in^2
1 atm	-do-	10.13×10^4 N/m^2	14.7 lbf/in^2
1 mm Hg	-do-	133.32 N/m^2	0.01934 lbf/in^2
10 poise	Viscosity	1 N s/m^2	2.089×10^{-3} lbf^{-3}ft^2
1 kgf m	Energy or work	9.81 J	0.1383 ft lbf
1 cal	Heat energy	4.1868 J	3.968×10^{-3} BTU
1 kW	Power	3.6×10^6 J	1.34 hp
1 kgf m/s	-do-	9.81 W	0.1383 ft lbf/s
1 hp	-do-	735.5 W	0.987 hp
1 kcal/hr	Heat flow rate	1.163 W	0.252 BTU/hr
1 kcal/m^2	Heat flow per unit area	1.163 W h/m^2	0.02712 BTU/ft^{2}hr °F
1 kcal/m^2 hr °C	Heat transfer coefficient	1.163 W/m^2 K	0.2048 BTU/ft^{2}hr °F
1 kcal/kg	Enthalpy	4.1868 kJ/kg	0.1145 BTU/lb
1 kcal/kg °C	Specific heat	4.1868 kJ/kg K	0.0635 BTU/lb °F
1 kcal/m hr °C	Thermal conductivity	1.163 W/m K	0.6723 BTU/ft hr °F

SCHMIDT NUMBERS OF GASES in dilute mixture with air†

GAS	Molecular Weight M	Schmidt Number Sc^*
Hydrogen	2.016	0.22
Methane	16.04	0.84
Ammonia	17.03	0.61
Steam	18.016	0.60
Nitrogen	28.02	0.98
Ethane	30.07	1.22
Oxygen	32.00	0.74
Methyl alcohol	32.04	1.00
Carbon dioxide	44.01	0.96
Propane	44.09	1.51
Ethyl alcohol	46.07	1.30
Acetone	58.08	1.60
Butane	58.12	1.77
Acetic acid	60.05	1.24
n-Propyl alcohol	60.09	1.55
Sulphur dioxide	64.06	1.28
Chlorine	70.90	1.42
Pentane	72.15	1.97
Methyl acetate	74.08	1.57
Ethyl ether	74.12	1.70
n-Butyl alcohol	74.12	1.88
Carbon disulphide	76.13	1.48
Benzene	78.11	1.71
Ethyl acetate	88.10	1.84
Toluene	92.13	1.86
Phosgene	98.92	1.65
n-Propyl acetate	102.13	1.97
Chlorobenzene	112.56	2.13
n-Octane	114.22	2.62
Napthalene	128.16	2.57
Carbon tetrachloride	153.84	2.13
Bromobenzol	157.02	1.97
Chloropicrin	164.39	2.13
Ethylene bromide	187.88	1.97

† Schmidt Numbers of gases in dilute mixture with air may be deduced from the following approximate relationship :

$$Sc^* \simeq 0.145 \, M^{0.556}$$

Schmidt number of a gas mixed with air in any proportion may be determined in the following way :

$$Sc = \frac{\nu}{\nu^*} Sc^*$$

where

ν, viscosity of the mixture, m^2/h

ν^*, viscosity of the component in a dilute mixture at the mixture temperature, m^2/h.

147

REFERENCES

Eckert, E. R. G. and Drake, R. M. (1959). *Heat and Mass Transfer*, McGraw-Hill.

Fishenden, M. and Saunders, O. A. (1957). *An Introduction to Heat Transfer*, Oxford.

Giedt, W. H. (1958). *Principles of Engineering Heat Transfer*, Van Nostrand.

Holman, P. J. (1963). *Heat Transfer*, McGraw-Hill.

Hughes, W. F. and Brighton, J. A. (1967). *Theory and Problems of Fluid Dynamics*, Schaum Publishing Co.

Jakob, M. (1949). *Heat Transfer*, Vol. I, Wiley.

Kern, D. Q. (1950). *Process Heat Transfer*, McGraw-Hill.

Kreith, F. (1958). *Principles of Heat Transfer*, International Textbook Co.

Kutateladze, S. S. and Borishankii, V. M. (1966). *A Concise Encyclopedia of Heat Transfer*, Pergamon Press.

McAdams, W. H. (1954). *Heat Transmission*, McGraw-Hill.

Rohsenow, W. M. and Choi, H. Y. (1961). *Heat, Mass and Momentum Transfer*, Prentice-Hall.

Schneider, P. J. (1957). *Conduction Heat Transfer*, Addison-Wesley Publishing Co.

Spalding, D. B. (1963). *Convective Mass Transfer : An Introduction*, Edward Arnold.

Streeter, V. L. (1962). *Fluid Mechanics*, McGraw-Hill.

Treybal, R. E. (1955). *Mass Transfer Operations*, McGraw-Hill.

Wrangham, D. A. (1961). *The Elements of Heat Flow*, Chatto and Windus.

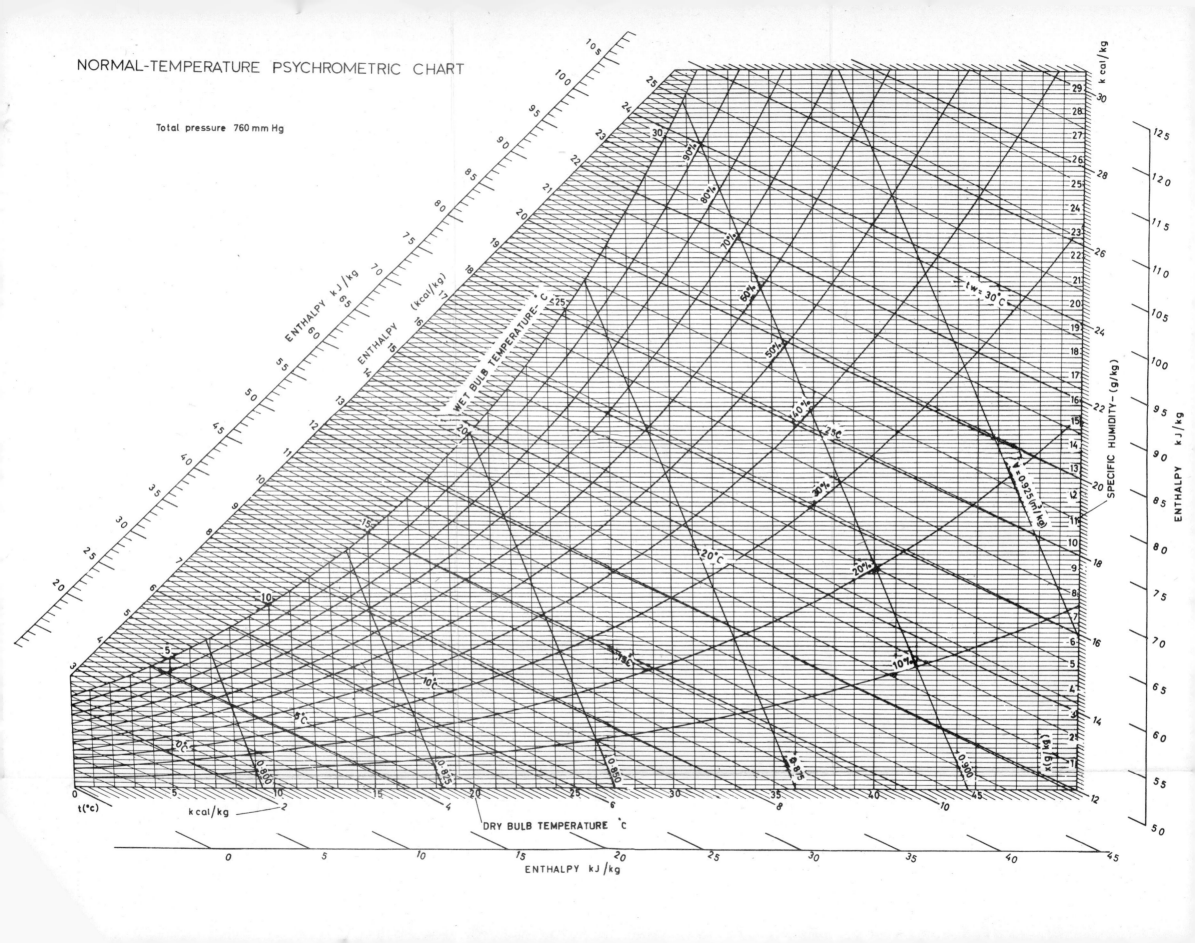

NORMAL-TEMPERATURE PSYCHROMETRIC CHART

Total pressure 760 mm Hg

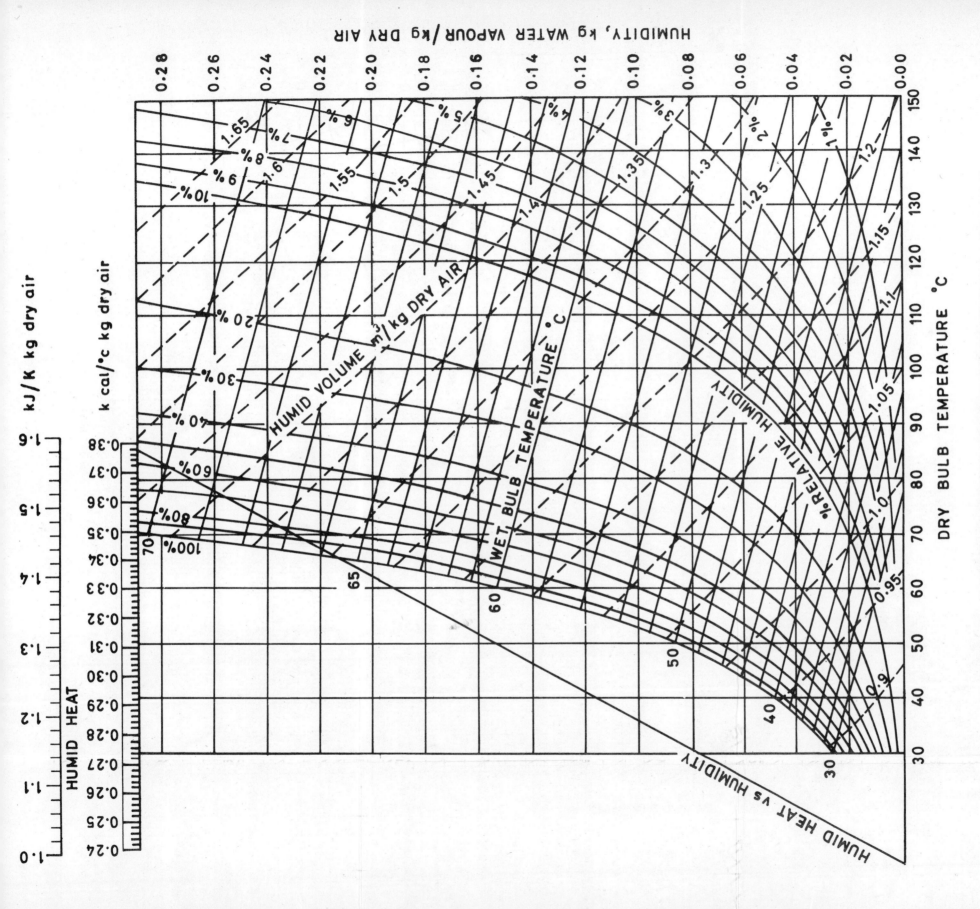

HIGH TEMPERATURE PSYCHROMETRIC CHART

(Courtesy : ATIRA, Ahmedabad)